Quality by Design for Electronics

Quality by Design for Electronics

Werner Fleischhammer

CHAPMAN & HALL
London · Glasgow · Weinheim · New York · Tokyo · Melbourne · Madras

Published by Chapman & Hall, 2–6 Boundary Row, London SE1 8HN, UK

Chapman & Hall, 2–6 Boundary Row, London SE1 8HN, UK

Blackie Academic & Professional, Wester Cleddens Road, Bishopbriggs, Glasgow G64 2NZ, UK

Chapman & Hall GmbH, Pappelallee 3, 69469 Weinheim, Germany

Chapman & Hall USA, 115 Fifth Avenue, New York, NY 10003, USA

Chapman & Hall Japan, ITP-Japan, Kyowa Building, 3F, 2-2-1 Hirakawacho, Chiyoda-ku, Tokyo 102, Japan

Chapman & Hall Australia, 102 Dodds Street, South Melbourne, Victoria 3205, Australia

Chapman & Hall India, R. Seshadri, 32 Second Main Road, CIT East, Madras 600 035, India

First edition 1996

Typeset by Concerto, Leighton Buzzard, Bedfordshire

Printed in Great Britain by Hartnolls Ltd, Bodmin, Cornwall

ISBN 0 412 56360 6

The terms 'he' and 'his' are not indicative of a specific gender and should be taken throughout to refer to both men and women.

A catalogue record for this book is available from the British Library

Printed on acid-free text paper, manufactured in accordance with ANSI/NISO Z39.48-1992 and ANSI/NISO Z39.48-1984 (Permanence of Paper).

Contents

Preface ix

1. Introduction **1**

2. Integrated quality assurance **9**
2.1 Quality culture 9
2.2 Quality assurance in the initial phase 11
2.3 Quality assurance in the project phase 12
2.4 Quality assurance during the design phase 15
2.5 Quality assurance in the prototype and early production phase 16
2.6 Quality assurance in the production phase 17
2.7 The way to lower cost 17

3. Joint evaluation of all components **21**
3.1 Evaluation, an important aspect of quality assurance 21
3.2 Evaluation survey and information system 27

4. Functional and electrical evaluation of digital ICs **35**
4.1 Logic function 38
4.2 Static characteristics 39
4.2.1 *Input/output characteristics* 40
4.2.2 *Transfer characteristics* 45
4.2.3 *Supply current characteristics* 48
4.2.4 *Special static characteristics* 49
4.3 Dynamic characteristics 52
4.3.1 *General considerations* 52
4.3.2 *Test set-up* 53
4.3.3 *Delay parameters* 53
4.3.4 *Transition time (rise and fall time)* 58
4.3.5 *Logic spikes and noise may increase delay* 60
4.4 Evaluation by automatic test systems 62
4.4.1 *The benefits of automated testing* 62
4.4.2 *Cost comparison* 66
4.5 Special electrical parameters 68
4.5.1 *Dynamic noise immunity* 68
4.5.2 *Threshold time* 70
4.5.3 *Dynamic power supply current and supply spikes* 72

4.5.4 *Ground bounce and bus contention* 73
4.5.5 *Capacitance of inputs and outputs* 75
4.5.6 *Asynchronous behaviour* 78
4.6 Evaluation of standard LSI and VLSI circuits 86
4.7 Electrical evaluation of asics 92
4.7.1 *Different categories of asics* 92
4.7.2 *Full custom asics* 92
4.7.3 *Semicustom asics: gate arrays and standard cells* 93
4.7.4 *User programmable asics* 111
4.8 Electrical evaluation of memories 114
4.9 Evaluation of ECL circuits 115
4.9.1 *Status of ECL technology* 115
4.9.2 *Evaluation tests on ECL circuits* 116
4.10 Evaluation of passive components, discrete semiconductors and analogue ICs 116
4.11 Subcontractors (OEM) 118
4.12 Final remarks on electrical evaluation 119

5. Reliability and environmental requirements 121
5.1 Reliability evaluation 121
5.2 Reliability tests 123
5.3 ESD sensitivity 124
5.3.1 *Physical basis* 124
5.3.2 *Measurement methods* 125
5.3.3 *Results* 127
5.4 Latch-up effect 130
5.5 Soft errors 131
5.6 Preconditioning (burn-in) 135
5.7 Wear-out 140
5.8 Technological evaluation 141
5.9 Mechanical characteristics 141
5.10 Customer audits at vendor's site 145
5.11 Cost and benefits of evaluation 145
5.11.1 *How to calculate the benefits of evaluation* 145
5.11.2 *Some real problems as examples* 151

6. Quality assurance in the production phase 153
6.1 General considerations 153
6.2 Quality management 154
6.2.1 *Failure management system* 154
6.2.2 *Manufacturing process control* 157
6.2.3 *Quality actions* 159

	6.2.4	*Test strategy*	160
6.3	Traceability		162
6.4	Incoming inspection		165
	6.4.1	*Management of incoming inspection tests*	165
	6.4.2	*How to choose the right testing hardware*	166
	6.4.3	*How to procure test software*	175
	6.4.4	*To what extent do components have to be tested?*	177
6.5	Board test strategy		179
	6.5.1	*Procedure for choosing the optimal test strategy*	179
	6.5.2	*Execution of board tests*	187
	6.5.3	*Design for testability*	191
	6.5.4	*Future trends*	191
6.6	Module test		193
6.7	Final or end test		194
	6.7.1	*Test procedure*	194
	6.7.2	*Detect board failures at board test, not at final test*	199
	6.7.3	*Optimal scheduling of subtests*	201
	6.7.4	*Test handler*	202
	6.7.5	*Use of an expert system*	203
	6.7.6	*Achieved cost reduction*	205
6.8	Service and maintenance		205
6.9	Failure analysis		209
	6.9.1	*Failure documentation*	209
	6.9.2	*Execution*	209
	6.9.3	*Process of analysis*	211
	6.9.4	*Interpretation of results*	213
6.10	Correlation with the vendor to reduce failure rate		216
6.11	Failure prevention		219
	6.11.1	*Statistical process control (SPC)*	220
	6.11.2	*Preconditioning (run-in)*	224
	6.11.3	*ESD prevention*	224
	6.11.4	*Use of human resources*	227
	6.11.5	*Permanent control and calibration of test equipment*	229
6.12	Process changes		233
	6.12.1	*Procedures for introducing process changes*	232
	6.12.2	*Process qualification and release*	233
	6.12.3	*Failure mode and effect analysis (FMEA)*	234
6.13	Achieved cost reduction		236

7. Design for quality, the key to good quality at low cost **241**

8. Specifications and standards as a basis for co-operation **247**

8.1 Purpose and procedure of specifying 247
8.2 General quality specification 247
8.3 Family specification 248
8.4 Part specification 248
8.5 Purchase contract 249
8.6 Agreement on the quality assurance (ship-to-stock contract) 249
8.7 ISO 9000 certification 249

Appendices

A: Propagation of signals and crosstalk on interconnection lines 253
B: Evaluation test program 265
C: Calculation of failure rate λ 267
D: Reliability questionnaire 269
E: Check-list for comparing ATE testers 279
F: Principles of test software generation for ICs of high complexity 285
G: Shielding effectiveness of a cable connection 293
H: Agreement on quality assurance between purchaser and supplier 297
I: General quality specification 301

Glossary **313**

Further reading **319**

Index **325**

Preface

This book concentrates on the quality of electronic products. Electronics in general, including semiconductor technology and software, has become the key technology for wide areas of industrial production. In nearly all expanding branches of industry electronics, especially digital electronics, is involved. And the spread of electronic technology has not yet come to an end. This rapid development, coupled with growing competition and the shorter innovation cycle, have caused economic problems which tend to have adverse effects on quality.

Therefore, good quality at low cost is a very attractive goal in industry today. The demand for better quality continues along with a demand for more studies in quality assurance. At the same time, many companies are experiencing a drop in profits just when better quality of their products is essential in order to survive against the competition.

There have been many proposals in the past to improve quality without increase in cost, or to reduce cost for quality assurance without loss of quality. This book tries to summarize the practical content of many of these proposals and to give some advice, above all to the designer and manufacturer of electronic devices. It mainly addresses practically minded engineers and managers. It is probably of less interest to pure scientists. The book covers all aspects of quality assurance of components used in electronic devices. Integrated circuits (ICs) are considered to be the most important components because the degree of integration is still rising. Therefore, the main focus is the application of these circuits in electronic devices. This book may be of some value to the designer of ICs as well, but no detailed theory of failure modes is given. The goal is to present methods on how to best use the benefits of quality assurance to improve product quality without increasing product cost. This book will show that it is profitable to involve quality engineers in the whole design process, starting with quality studies in parallel with first design steps and ending with periodical quality reviews together with the vendors. Likewise, open co-operation with the vendors in all stages of design will reduce the cost of quality assurance considerably.

The impetus to write this book came from many discussions which the author had with several manufacturers of electronic devices, mainly computers, and with many vendors of electronic components from various countries. The main topics of these conversations involved finding a way of achieving an optimal compromise between quality and cost, and a way of balancing the partly conflicting interests of customer and vendor of components. The experiences, good and bad, which the author gained during this task are laid down in this book. However, it should be mentioned that, in this book, the problems are viewed from the eyes of a user more than from the eyes of a vendor of components. In this way, it may be a useful complement to the literature on the subject where quality problems are mostly seen from the viewpoint of the manufacturer of components. Aspects of software quality are not included in this book.

The author thanks everyone, vendors and users of ICs who assisted him in writing this book for all their fruitful discussions and valuable advice. Special thanks are due to Dr N. Lieske, director of quality assurance, for his input on joint qualification, preconditioning and ESD prevention and to Mr M. Kuchler, director of quality control, Siemens Nixdorf AG., Augsburg, Member of ZVEI ('Association of Information and Communication Technology') for his contribution to CAM, as well as to Mr F. Gelzer and Mr G. Gelfort for their help in preparing chip photos.

Last but not least I have to express my personal thanks to Mr Charles Gardiner and to Mr Philip Walters for their kind linguistic corrections, and to Jenny Lawson and her crew at First Class for editing this book.

Werner Fleischammer
Brunnthal, 1995

1

Introduction

When valves were the active elements in electronic devices, their lifetime was so much less than the lifetime of all other components (like resistors or capacitors), that the quality of valves determined the quality of the device.

So, when valves were replaced by transistors (1955–1960), many people thought that quality assurance no longer had any meaning because the lifetime of transistors was assumed to be the same as passive components, nearly indefinite. But soon people became aware that transistors also have a limited life.

Since then, two phenomena have steadily increased the importance of quality assurance: the demand for more performance and the demand for better quality, both at lower cost. The traditional price per performance ratio expanded to the relation between price, performance and quality.

Firstly, the complexity of electronic devices and the number of components increased. This is demonstrated by Figs 1.1–1.3, which show the increase in gate count for microprocessors, asics (application specific integrated circuits), and RAMs (random access memories); and also the gate density in gates/mm^2 or the structural dimension (channel length) in μm. This rise, which shows no flattening until now, is often called Moore's law.

The increase in gate count was more than the increase in gate density because the dimensions of the chip increased too. The intensive usage of these very large-scale integrated circuits made possible the rise in performance of the electronic devices seen today.

Table 1.1 shows the rise in performance of computers, mainly office computers or small mainframes in the last few years, and the prediction for the near future. The gate count and the performance of computer mainframes increased too, whereas their dimensions remained unchanged or decreased. Mainframes are chosen as examples of advanced devices, but similar figures can be derived for other kinds of electronic devices such as telecommunication circuits (Bursky, 1993; Tanaka, 1992).

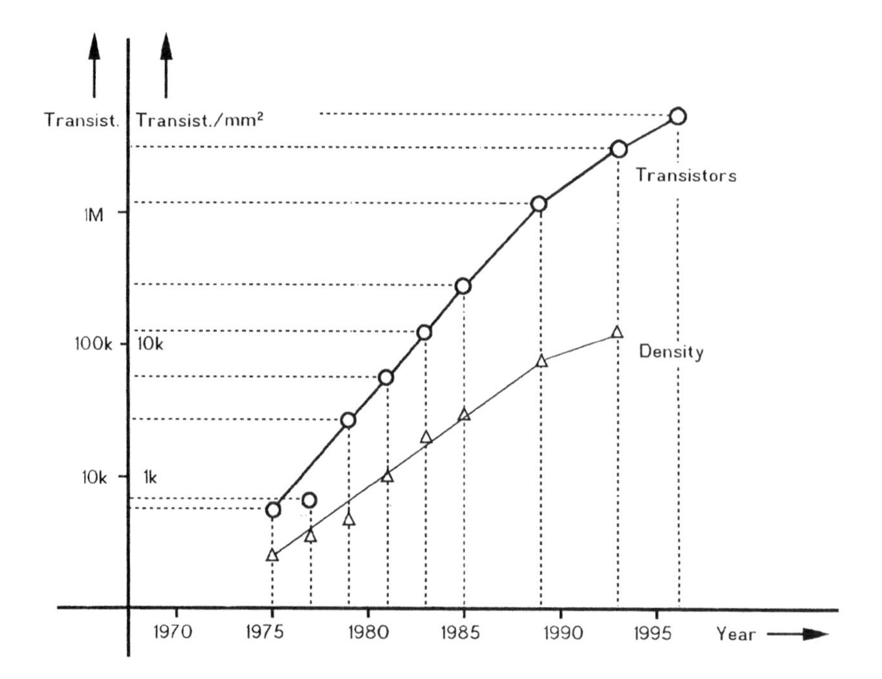

Year[1]	*Reftyp*[2]	*Transistors*[3]	*Density*[4]
1975	8 080	5 500	260
1977	8 085	6 000	350
1979	8 086	29 000	450
1981	80 186	55 000	1 000
1983	80 286	130 000	2 100
1985	80 386	275 000	2 800
1988	80 486	1 200 000	7 300
1992	Pentium	3 100 000	
1996	P6	5 500 000	≈15 000

[1] Year means the year of final samples.
[2] Intel was used as reftyp because it was most popular. Other vendors offer comparable circuits.
[3] Transistor count was from vendor's data.
[4] Maximum density in transistors/mm^2 was from chip photographs. Overall density is about 60% of maximum.

Figure 1.1 The rising complexity of microprocessors: by a factor of ≈10 every six years.

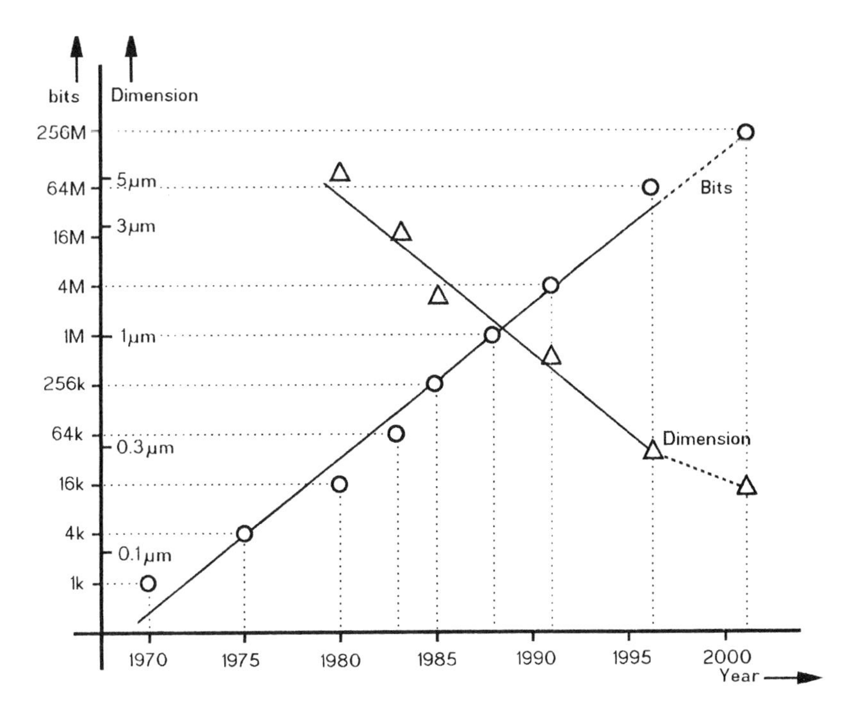

Figure 1.2 DRAMs lead the way into high complexity and small structural dimensions. The time-scale is based on shipment of more than 10 million pieces per month.

Table 1.1 Increase of performance and gate count of office computers

Year	*RPF*[1]	*Gates*[2]	*LSIs*	*Gates/LSI*[3]	*Components/PB*[4]
1990	3	260k	8	30k + 40kb	70
1992	8	480k	5	120k + 50kb	70
1995	30	1.0M	4	300k + 1Mb	80

1 Relative performance. The author tried to find a measure to compare the performance coming from different sources.

2 Gate count estimated in two input nand gates. For asics a usage of 60–75% of maximum gates was assumed.

3 Maximum available logic gates plus RAM cells per asic.

4 Number of active ICs only per board. Each office computer was assembled on one board only.

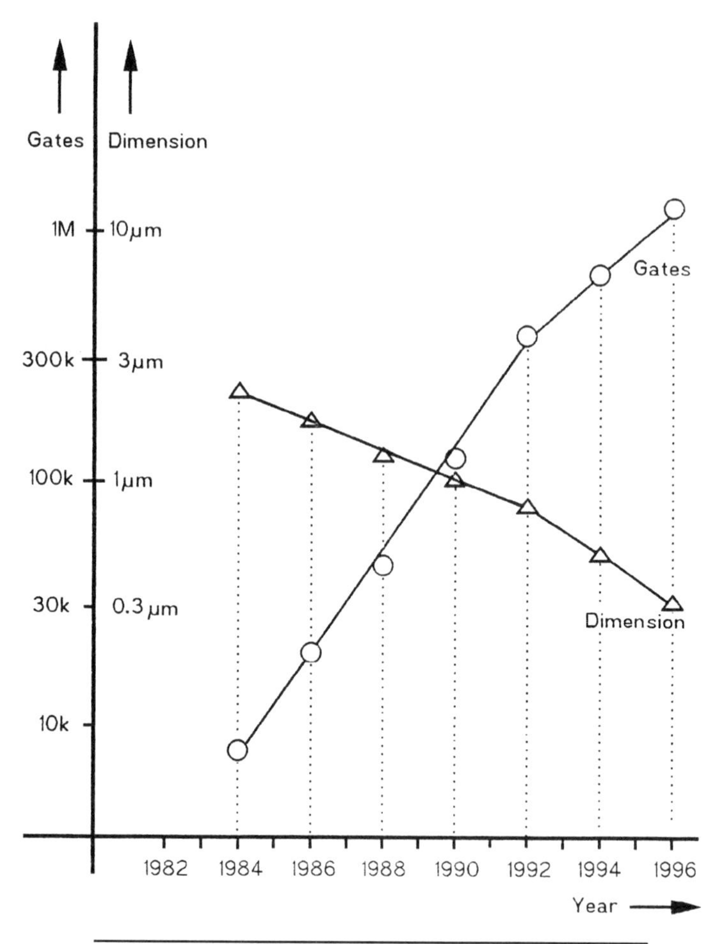

Year	*Gates*	*Dim. (μm)*	*Ref. vend.*[1]
1984	8 000	2.5	Fujitsu
1986	20 000	1.8	Fujitsu
1988	45 000	1.2	Siemens
1990	115 000	1.0	LSI-Logic
1992	318 000	0.8	Motorola
1994	500 000	0.5	LSI-Logic
1996	1 400 000	0.25	IBM

[1] Reference vendors are examples only.

Figure 1.3 Asics are coming up.

The demands for more quality were also increasing. Figure 1.4 shows the mean time between failure in the field for mainframes. The failure rate had been decreasing in spite of the increasing performance and complexity.

Both, the increase in complexity and the demands for more quality of electronic devices, led to more demands on the quality of the single components and production methods and to more attention to quality management. This implies not only the incoming inspection of all components, usually a 100% test of key parameters, but also the thorough examination and qualification of these components on a sample basis. Qualification tests include measurements of all electrical and mechanical properties, and also include environmental and life tests. In the same sense, a qualification of all production methods, as well as intermediate and final tests of the products, are part of quality assurance.

But better quality has its price. Because of the considerable effort required, the question arises as to whether the rising costs associated with the steps listed above can be reduced. The answer to this question depends on the requirements specific to the application. Several strategies have been established for different products (Table 1.2).

Table 1.2 How much testing is necessary?

Product	*Tests*	*Production*	*Quality*
Consumer	Few	Cheap	Moderate
Military	Many	Expensive	Excellent
Industrial	Some	Rational	Sufficient
Goal	Few	Rational	Excellent

For military products, optimum quality is most important, and the cost of quality plays a secondary role. The same is true for life-sensitive products. For consumer products, low cost is a primary factor, and a limited lifetime is often taken into account. Industrial products lie somewhere in between.

The final goal should be to achieve good quality at low cost by rationalized production and a minimum of testing. Let us now consider a three-pronged method to reach this goal.

1. The integration of quality assurance into the design cycle of electronic devices by a fundamental change in management

philosophy is essential. Quality assurance has to become a major design goal and has to be considered as early as possible by all design engineers at all levels. A quality culture has to be established. This is covered in Chapter 2.

2. The joint qualification of components is needed to minimize the effort expended in qualifying and evaluating by both vendor and customer in a true partnership. This is explained in Chapter 3. Because this goal is not yet commonly understood, it is dealt with more explicitly in this book: the details of electrical evaluation in Chapter 4, and Chapter 5 listing the reliability and environmental requirements. Some people still doubt that a profit can be

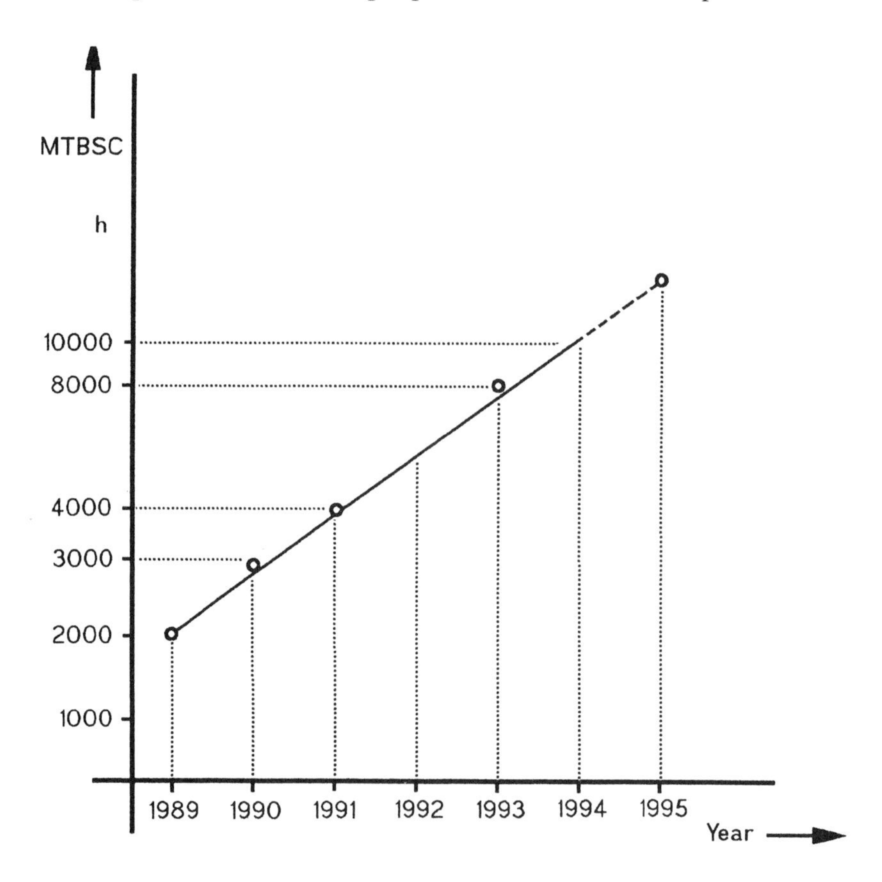

Figure 1.4 The answer to customers' requirements: the mean time between service call (MTBSC) increases.

achieved when there is a thorough and complete evaluation of all components; but potential failures which are detected before they occur, produce the least cost. How to calculate the benefits of evaluation is shown at the end of Chapter 5. In this respect, Chapters 3 to 5 have to be seen as a whole.

3. The reduction of cost for quality assurance in the production phase without loss of quality by open co-operation between vendor and user of electronic components. This implies the reduction of inspection on incoming deliveries with the goal of abolishing incoming inspection completely. It requires the building up of trust as a precondition and periodical quality reviews during the whole production period. These objectives will be the primary contents of Chapter 6.

The results of Chapters 3 to 6 are summarized in Chapter 7, which shows how the final goal, high quality at low cost, can be achieved step by step. Whereas co-operation between vendor and customer in quality assurance has become well established, at least for major customers, similar co-operation in the evaluation stage is still in its

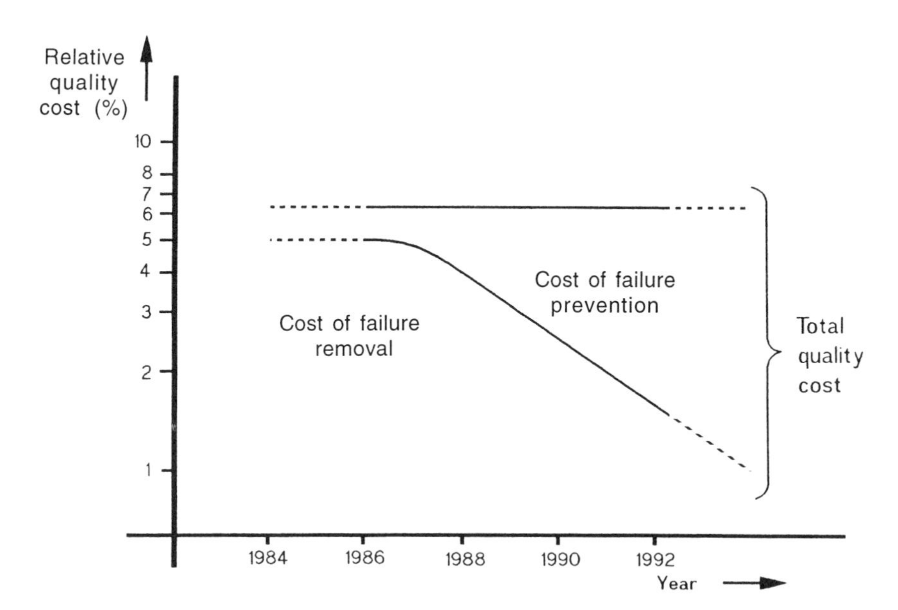

Figure 1.5 Better quality without rising costs: the shift from repair to prevention.

infancy. Therefore more weight has been given to Chapter 3.

It is extremely difficult to obtain realistic and accurate figures of the cost of quality. Managers always tend to give positive answers, perhaps only admitting that it was worse in the past. So the changes in the relative quality cost shown in Fig. 1.5 have to be taken as an estimate derived from few inputs and should be seen in conjunction with Fig. 1.4.

While the quality improved, the total cost for quality remained constant at about 5–15% of the total production cost, centred at 6–7%. The cost for failure removal and repair decreased considerably, whereas the cost for failure prevention increased at the same time. But these figures can only be an example, in each particular case detailed cost considerations have to be made to arrive at actual values for a specific situation. How this can be done is explained in Chapter 6.

In Chapter 8, a realistic example is given of how the co-operation between vendor and customer can be specified in a contract.

2

Integrated quality assurance

2.1 QUALITY CULTURE

Quality assurance in the past was often an appendix to manufacturing. Its main task was to assure the quality of outgoing products by making measurements on attributive samples. These included accelerated life tests and tests under environmental stress like shock or aggressive atmosphere. It was up to management to interpret the results and decide whether corrective changes had to be made. This situation could never be very effective, because it resulted in curing bad quality instead of preventing it.

In order to produce really high quality products without excessive cost, it is of little use to append quality assurance to the end of the design cycle. It has to be taken into consideration from the earliest point, starting during the project phase when engineering is dealing with block schematics, with performance requirements and with development costs of the planned device. At this time, when the design is nothing more than a faint idea and quite open to changes, quality engineers have to become involved (Table 2.1) because quality has to be designed not only into components but also into complete electronic devices. The integration of quality assurance into the design cycle is an important management task.

The design process too has to be systemized and structured with built-in quality reviews between successive design steps. In these reviews, the design has to be analysed systematically and critically for potential quality risks. This does not mean restricting the initiative of the design engineers, but creating an awareness of the significance of quality not only as a general goal but in every step of the design process. Each member of the company has always to bear in mind the sequence: 'Good work – contented customers – new orders.' It is upper management's task to encourage this. Not only is a demand for better quality necessary but so is a correct assessment and valuation of it.

Table 2.1 Integration of quality assurance into the design cycle

Pre-evaluation stage	
Engineering	First design idea, block schematics List of required circuit types
Quality group	Study of vendors' datasheets Study of testability Study of packaging Study of availability Study of expectable quality
Management	Decision to start detailed design
Design and evaluation stage	
Engineering	Detailed design and simulation
Quality group	Electrical evaluation Static tests Dynamic test Special tests Functional tests Technological inspection Reliability tests Packaging and manufacturing tests Preparation of loadboards and test programmes for incoming inspection Define optimal test concept Prepare specifications
Management	Decision to start production
Production phase	
Quality group	Control incoming inspection and production tests Failure analysis Corrective loops Failure statistics Cost reduction Ship-to-stock
Management	Periodical quality reviews

An understanding of the importance of quality in the end product is a prerequisite to competing in the market. This feeling has to be transferred from the quality engineers to the design engineers. Outstanding electrical performance and a competitive price are both necessary, as well as quality which satisfies the customer. That means a total quality culture has to be established. Quality seminars have to be held to attain a consciousness of quality at all levels. Both managers and engineers must be invited to take part in workshops where quality problems can be openly discussed with electronic design engineers, and where quality engineers advise the design management. But this also means that the quality engineers must be convinced of the importance of their job and of their responsibility for the success of their company. The executive management has to demonstrate its interest in quality and formulate it as a strategic goal of high priority (Ishikawa, 1985).

2.2 QUALITY ASSURANCE IN THE INITIAL PHASE

The more demanding electronic products become, the more preparatory work has to be undertaken at the beginning of a new project. More than two thirds of all quality problems have their roots far beyond the start of production. The decisions which influence the quality of hardware and software products are made in the early design and development stage, where points such as care of manufacture, automation, testability, purchase of components without difficulty and maintenance are considered. Every later update or improvement becomes more costly and time consuming. Therefore the centrepoint of modern quality assurance is shifted more and more towards the initial planning and design stage. It is there that quantitative quality indications and objective quality standards have to be put in place, and definite quality design goals have to be agreed upon and pursued systematically.

During planning, decisions will be made by a planning team concerning size, performance and the rough overall structure of this new device. These include not only the price and performance but also decisions on the kind of components, their technology, the manufacturing tools to be used and a proposal for the timeframe for the whole development process. The quality mangement representative within this team will have to contribute the fundamental quality considerations and introduce aspects of quality into new design projects.

The following practical example from the world of computer mainframes demonstrates the importance of this contribution.

Increasing integration density increases the quality in general because of the lower part count which can be achieved. Using high-speed technology, like ECL, will increase performance, but its higher power consumption will eventually reduce the degree of integration. The same performance can be reached using lower speed circuits by parallel computing of several processors. This increases the amount of logic calculated in gate functions but, because of the higher integration density, the part count may be lower.

Now, parallel computing needs special software which causes additional cost, but it increases MTBF by its capability of automatic reconfiguration which allows continued function in spite of a failure. So, a trade-off between the different influences has to be made.

Figure 2.1 shows the estimated board count and relative performance of this multiprocessor system, quoted here as an example, as a function of the number of processors for ECL and CMOS technology. It also shows the prediction of MTBF determined by the quality group. Whereas at the beginning of the discussion an ECL double processor was anticipated, after the discussion a quadruple CMOS processor was chosen because of its higher MTBF and smaller size at the same performance.

This does not mean that quality mangement takes the decisions, but it does have a duty to show up the quality risks and trade-offs of different alternatives. The cheapest failure is always the failure which does not occur. So, it is worthwhile thinking about quality prior to any design activity.

2.3 QUALITY ASSURANCE IN THE PROJECT PHASE

After having made a first sketch of the design project, the design engineers have to give a preliminary list of all new components which they have in mind and of new production methods which might become necessary. Then the quality management has to take the following pre-evaluation and studies into consideration.

1. **Study of vendors' datasheets of all new circuits**
   Are they complete and fully understood or are there contradictions or ambiguities? Any dubious statement in a datasheet should be made clear by a written comment by the vendor. This avoids many problems which can occur later during evaluation or even during application of the circuit. The crystal inputs of some early microprocessors can be cited here as an example. The requirements on the crystals were almost always unspecified by the vendor.

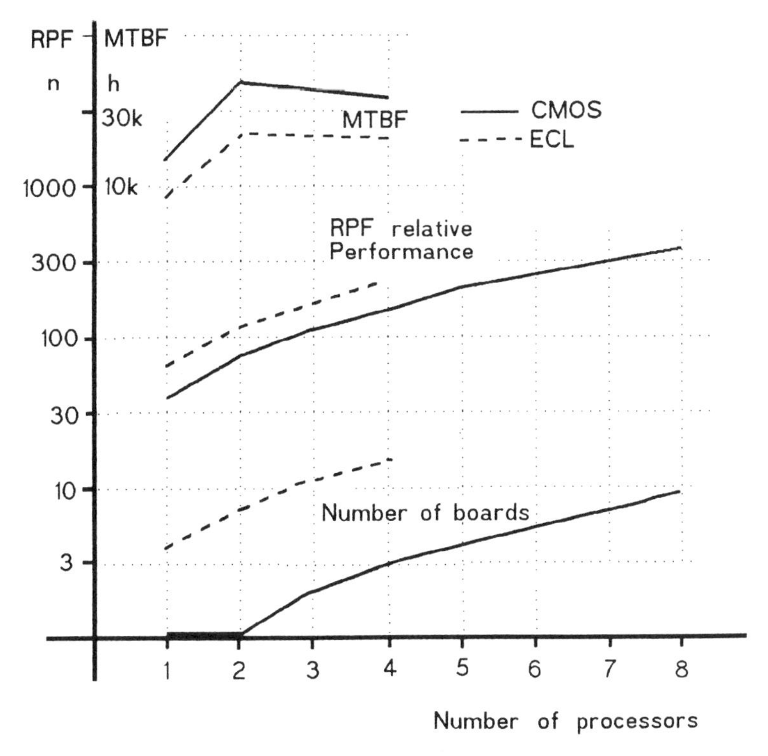

Figure 2.1 The trade-off between speed and integration: quality is part of the decision.

2. **Study of testability**
   Are all new components testable with available testers? Are excessive investments to generate test programmes expected? This is of great importance for more complex components such as digital and analogue integrated circuits (ICs). Modern asics with extremly high pin counts of more than 500 pins may cause problems at the test stage.
3. **Study of availability**
   Are all vendors qualified as reliable suppliers? Are second sources available or expected? Design engineers may find a sophisticated product from a new and unknown vendor fits their needs exactly. Should the vendor later cease delivery of this part, the consequences will be catastrophic. It will cause a delay or stop production because of necessary redesigns. A solution for such a

case could be to oblige the design group to prepare a lifebelt, i.e. a less optimal alternative in case of emergency. (A detailed discussion of this point for asics is given later.)

4. **Study of manufacturability**
   Are new package types planned which may cause problems in handling and solderability? In the past, a change from DIP to surface-mounted packages or from bipolar to MOS technology had a great impact on the whole production. Similar problems will arise if, for instance, caseless packages or something similar are introduced.
5. **Study of reliability**
   Which time-consuming reliability tests are expected? Are all circuits approved (UL, VDE, etc.)? What risks will arise due to product liability legislation?
6. **Study of design tools**
   Are new design tools, new libraries or an upgrade of existing libraries necessary? Simulation is absolutely recommended for asics but it is also very useful to simulate a complete electronic device. This may save having to search for failures in prototypes, or building a prototype at all, and so reduce time to market. In terms of quality, it will reduce reworks considerably.

   Figure 2.2 shows how simulation helped to detect failures during a mainframe design containing about 250 000 gate functions. More than 95% of all design errors were detected by simulation, prior to design release. No redesigns of asics were necessary and only a few changes had been necessary after power-on of the prototype. The forecasted timeframe from beginning of design until start of delivery could be reduced considerably. Although the investment and the effort necessary to simulate such rather large logic complexes were enormous, the result proved that it is worthwhile to do so.

   To prepare simulation software is the business of engineering management, but it is the quality manager who has to take care that the software from a new asic vendor fits into the established simulation landscape.

After completion of these studies, a management decision has to be made as to whether the planned design is reasonable or should be more conservative, using less advanced circuits, or should be scrapped altogether. All relevant groups – marketing, engineering, quality assurance, manufacturing, testing and maintenance – have to take part in this decision.

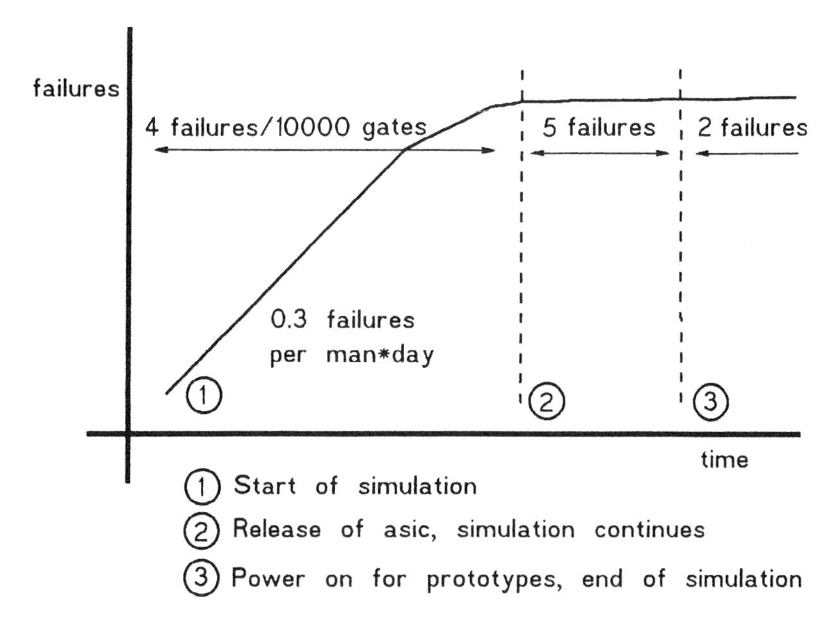

Figure 2.2 Simulation improves quality.

In the meeting, the quality group has to highlight the quality risks and costs in as much detail as possible. It should be noted that the decision has to be made with unanimous consent of all parties involved.

2.4 QUALITY ASSURANCE DURING THE DESIGN PHASE

If this management decision is positive, the design phase will follow. Engineers will detail logic design and make simulations. In this phase, the quality group is responsible for the following.

1. The procurement and the electrical evaluatuion of all new components:

   - static characteristics of inputs and outputs;
   - functional tests;
   - dynamic characteristics (delay, rise time, frequency);
   - special tests like noise immunity, ESD immunity, capacitance, asynchronous behaviour.

(All these tests have to be made under all environmental conditions of temperature, supply voltage, etc.)

Parameters which are insufficiently specified by the vendor have to be measured with special care. The behaviour of a sequential circuit when driven by asynchronous signals is one example of possibly missing data. This is discussed in detail in Chapter 4.

2. Starting negotiations with the vendor's design group to obtain an evaluation data package and to discuss it with the circuit designers. One major task is to establish personal contacts between the actual designers of both parties, so problems occurring later can be solved more easily, without formalities and in close co-operation.
3. Preparing test facilities, test programs and loadboards. Especially in case of asics, the procedure on exchanging test patterns has to be made clear with the vendors.
4. Carrying out life tests and other tests to assure quality, as well as doing some technological inspection.
5. Testing handling, solderability, inflammability, etc. in order to provide all necessary data to the manufacturing people.
6. Preparing specifications and agreeing them with the vendors.

All this is described in detail in Chapters 3–5. If all these steps are completed satisfactorily, then a second management decision must be made on whether to build prototypes and start production, or to do some redesigning first. In making this decision, the above results have to be taken very carefully into consideration. This is the last point in the timeframe where details of performance, cost, market demand and quality can be adjusted.

2.5 QUALITY ASSURANCE IN THE PROTOTYPE AND EARLY PRODUCTION PHASE

The next major task of the quality group is to solve problems which may arise during the prototype and early production phase. These may be excessive defect rates at incoming inspection or in later board, module or final tests, packaging problems during manufacturing, etc. Close co-operation with the vendors is necessary and the quality group's function is to provide a link. The main objective in this phase is to reduce production costs and the cost of quality assurance, without reducing quality. Failure analysis has to be done and failure statistics have to be set up and discussed with the vendors, bearing in mind the final goal of coming to a ship-to-stock agreement. The consequences of possible redesigns, done by the vendors on their components, have to be observed. Not only component failures but also failures caused by production have to be included in a corrective loop.

Total quality affects the manufacturer of electronic devices not only as a customer of components but also as a vendor of devices. In the future the quality of the end product may become more customer-driven than is the case today.

2.6 QUALITY ASSURANCE IN THE PRODUCTION PHASE

Once the production is running steadily, the design group will usually withdraw itself from the product and turn to new designs. It is now up to the quality group to be responsible for undisturbed production and the prevention of any standstill. Proper replacements have to be found in time for components which are no longer available from a vendor. If the quality of the final products grows worse then the quality group has to find the reason and take counter-measures. So the quality group is involved with a product from the pre-design stage until its production is discontinued.

This chapter shows that the integration of quality assurance into the complete design and production of electronic devices is essentially a management problem. But besides this, there are many detailed technical problems which are dealt with in Chapters 3–6.

Of course, the ideal procedure sketched above has to be modified in the real world, but the given procedure could act as a checklist for design flow, as shown in Table 2.1.

2.7 THE WAY TO LOWER COST

The goal of good quality at low cost requires perseverance and patience. There might be no sudden success and it takes time to come to the hoped-for result. First of all, a carefully thought out scheme to achieve the goal step by step must be put in place and then persistently followed, even if the cost rises initially.

The cost of failure detection, and repair, increases with the complexity of the test object. Therefore a shift of the test effort from incoming inspection to final test or even to maintenance will by no means reduce costs. On the contrary, the cheapest way to achieve good quality is to detect all failures at the earliest stage of the production flow (Table 2.2).

Table 2.2 Cost of failure detection depends on production step

Location	*Production step*	*Relative cost*
Manufacturer of ICs	Incoming inspection of raw materials	
	Wafer test	
	Outgoing test	
Manufacturer of devices	Incoming inspection of circuits	1
	Board and module test	10
	Final test	100
Customer	Detection of failure during maintenance	1000

1. The more complex the test object, the more expensive is failure detection.
2. The earlier the failure is detected, the lower is the cost for repair.
3. Therefore, the more complex the end product, the more profitable is early failure detection.

The manufacturer of ICs and the manufacturer of electronic devices form a quality team.

Tests have to be performed after each production step. It has to be noted, however, that the optimal number of test steps depends strongly on the production volume. The values given in the third row of Table 2.2 are rough estimates for a typical case and should be verified as described later in Chapter 6. In the worst case, for instance for a warranty repair at the customer's site in a foreign country, this ratio may be much worse.

However, this is only the first step. It will improve quality but not necessarily reduce costs. This can be seen from Figs 1.4 and 1.5. While the quality improved, the costs remained constant, but a shift from failure removal to failure prevention cost can be observed. Therefore the next steps, producing a detailed failure analysis of all faulty components and subsequent measures to prevent the occurrence of these failures in future, are absolutely necessary. It is of no use only replacing the faulty parts in order to get the product running, although it may be tempting to do so. The vendor and the customer have to correlate their effort to reduce the defect rate at incoming inspection by a true partnership. This is a necessary condition for a gradual reduction and finally abolition of testing. Not only component failures but also production failures have to be reduced by similar methods. The curtailment of board tests and final tests follows the abolition of incoming inspection. The objective is that the components will be delivered from the vendor to the customer's automatic placement machine, mounted and soldered on the boards, assembled automatically and tested by self-test. It is a long and tedious journey, but when you have arrived, then the goal, good quality at low cost, has been achieved. All this is explained in detail in Chapters 6 and 7.

But this ideal will not last. New, more advanced components will appear on the market and new production methods will be introduced. This could cause severe problems because all test equipment has been discarded and there are no skilled test personnel. For this reason, a continuous and fast-reacting quality monitoring system and an experienced troubleshooting team are essential – as is a complete and thorough evaluation of all new components and production procedures.

The first step on the way to lower cost, a considerable reduction of incoming inspection, has been reached by many companies by joint efforts between vendors and customers. The reduction of board tests by faultless production has made good progress recently and the thinning of maintenance and service in course of an improved MTBSC (mean time between service calls) is on the way. So it seems that quality assurance in the production phase is reaching its goal. It follows that the significance of quality assurance in the design phase will increase. This is the topic of the next chapter.

3

Joint evaluation of all components

3.1 EVALUATION, AN IMPORTANT ASPECT OF QUALITY ASSURANCE

As failure detection loses its prevailing importance, failure prevention comes into vogue, as stated in Table 3.1. The first step in failure prevention is the thorough evaluation of all new components. Evaluation means extended tests on a few samples of a component type as a first step towards its qualification. Due to the high complexity of modern electronic systems, high values of MTBF have to be realized. Therefore the importance of the qualification of electronic components is growing. This results from the fact that quality can only be improved when it is designed into the product. Therefore the goal for vendor and customer must be a joint qualification. The objective is, on the one hand, to avoid any duplication of work. The results of all characterization tests done by the vendor should be made available to the customer. The customer, on the other hand, should help the vendor to improve the components by transferring all the knowledge gained by evaluation and application back to the vendor. So a joint evaluation is indispensable to improving quality at a reasonable cost.

It is strongly recommended that as much information as possible is obtained from the vendor, even if no joint evaluation procedure has been formally installed. Much of the evaluation data, as listed in the next chapters, is available at the vendor's site, but mostly only within the engineering or quality department. The link to the customer is through sales people who are less informed on detailed technical data and whose main duty is often to shield engineering from customers' questions. Persistence is needed to break through this barrier.

Vendors are often hesitant about giving more than datasheet values. They avoid giving out information which cannot be guaranteed and which is subject to change. Of course, they are afraid of incurring any liability. The customer has to convince the vendor that all evaluation data requested will not be used against the vendor but will only be

Table 3.1 Failure prevention, a general trend in quality assurance

Failure detection loses its prevailing importance
Incoming inspection
Board test
Module test
End test
Provident maintenance
Failure prevention gains rising importance
Evaluation of all new components (advanced technology)
Evaluation of all new production processes (miniaturization)
Co-operation with vendors to achieve low defect rates
Process control methods to reduce manufacturing faults

Note: This is not a revolution but an evolutionary process.

used as an input to the customer's design department in order to avoid inappropriate application of his components. A clear and well-defined distinction has to be formulated between data which the vendor has to guarantee by purchase contract, and informative data without any liability for the vendor. If the customer should detect any shortcomings in these data, which may prevent him from using the components, then the vendor must have complete confidence that the customer will discuss this openly with him. This is also the way to a joint qualification of electrical properties.

This applies not only to evaluation measurements carried out by the vendor, but also to the vendor's design data. Today, for instance, most vendors model all newly designed circuits for spice simulation, or something equivalent, and it should be possible for a customer to obtain the models and their parameters from the vendor. Typical and worst-case static characteristics and delay values can then be calculated.

Many vendors are well prepared to respond to their customer's request and have evaluation data packages available. In Table 3.2, an example is given of the minimal amount of data which should be contained in a data package. These packages are necessarily incomplete because each customer has his or her own special ideas and requirements. The customers will analyse these data and, if the product does not meet their requirements, a meeting will be arranged with the vendor to discuss and allocate additional activities. These may include further measurements, done either by the vendor or by the customer (Chapters 4 and 5), or there may be other agreements like

Table 3.2 Contents of a data package

General data

Reason for qualification:

( ) New product release
( ) Major change
Device type and name:
Data code of tested samples:

Electrical characterization

Chip schematics on gate level/simulation model;
Chip schematics on transistor level/spice model;
Preliminary datasheet/database including test conditions;

Design characterization (Lab-data)

Input characteristics $I_{In} = f(V_{In})$;
Output high characteristics $V_{OH} = f(I_{OH})$;
Output low characteristics $V_{OL} = f(I_{OL})$;
Transfer characteristics $V_{Out} = f(V_{In})$;
Supply currents $I_{CCH} = f(V_{CC})$, $I_{CCL} = f(V_{CC})$;
Output for V_{CC}=0 V $I_{Off} = f(V_{Out})$;
Simultaneous switching $V_{OL} = f(t)$;
Slowly rising (~500 ns) input $V_{Out} = f(t)$;
Supply current for high frequency $I_{CC} = f(f)$;
Input and output capacitances;
Dynamic noise immunity and threshold time:

Product characterization (ATE data)

V_{In} at $I_{In} = -18$ mA	Clamp voltage
I_{Il} at $V_{In} = 0$ V	
I_{Il} at $V_{In} = 0.8$ V	Datasheet parameter
I_{Ih} at $V_{Ih} = 5.5$ V	Break through
I_{Ih} at $V_{Ih} = 2$ V	Datasheet parameter

. . . .
. . . .

T_{pdhl} at $T_A = -40°C$ to $+125°C$

. . . .
. . . .

and so on

Technology summary: See Appendix D

Reliability tests: See Appendix D

Manufacturability:

ship-to-stock contracts or quality specifications (Chapter 8). Figure 3.1 gives an overview of the interrelations between vendor and customer in a qualification procedure. Joint qualification is not just a slogan but a key procedure to reduce the cost for qualification.

Independent laboratories – sometimes called test labs or test houses – may be helpful to the customer in checking the vendor's data or carrying out further evaluation tests. This may bring cost advantages or at least make the costs for evaluation more calculable for the customer. Care is needed to select a test laboratory that is appropriate (Flaherty, 1993). Some specialize, in electrical or reliability tests or in physical failure analysis. Contact several to evaluate their technical expertise and capabilities. To avoid disappointment take a two-stepped approach. At first, use the test lab to unburden you from time-consuming test work.

- The test lab should have the capabilities in equipment and know-how to perform all evaluation tests which you consider necessary.
- The test lab should be able to assess the results and locate weak points.

If this works out successfully, then you may decide to take the next step of setting up a long-term technical partnership with this test lab. Some companies see this as a reliable way of securing emergency test capability after abolishing incoming inspection (Chapter 7).

- The test lab should have the technical expertise to vindicate poor or unacceptable results in discussions with your manufacturer.
- The test lab should be willing to assist you in analysing later application problems on the evaluated components.

But the main point in this partnership is that the test lab engineers exchange with your engineers all experience and knowledge gained on the behaviour of new technologies during evaluation. Never become totally dependent on a foreign institution.

Either way, it is imperative to conclude a contract with the test lab in which all duties are specified exactly, as later extensions will usually increase costs considerably. If you opt for the second step then an even more determinative agreement must guarantee the allocation of standby test capability and willingness to solve sudden problems in a predetermined time. But remember, test labs will never take over the responsibility for the quality of the customer's end products.

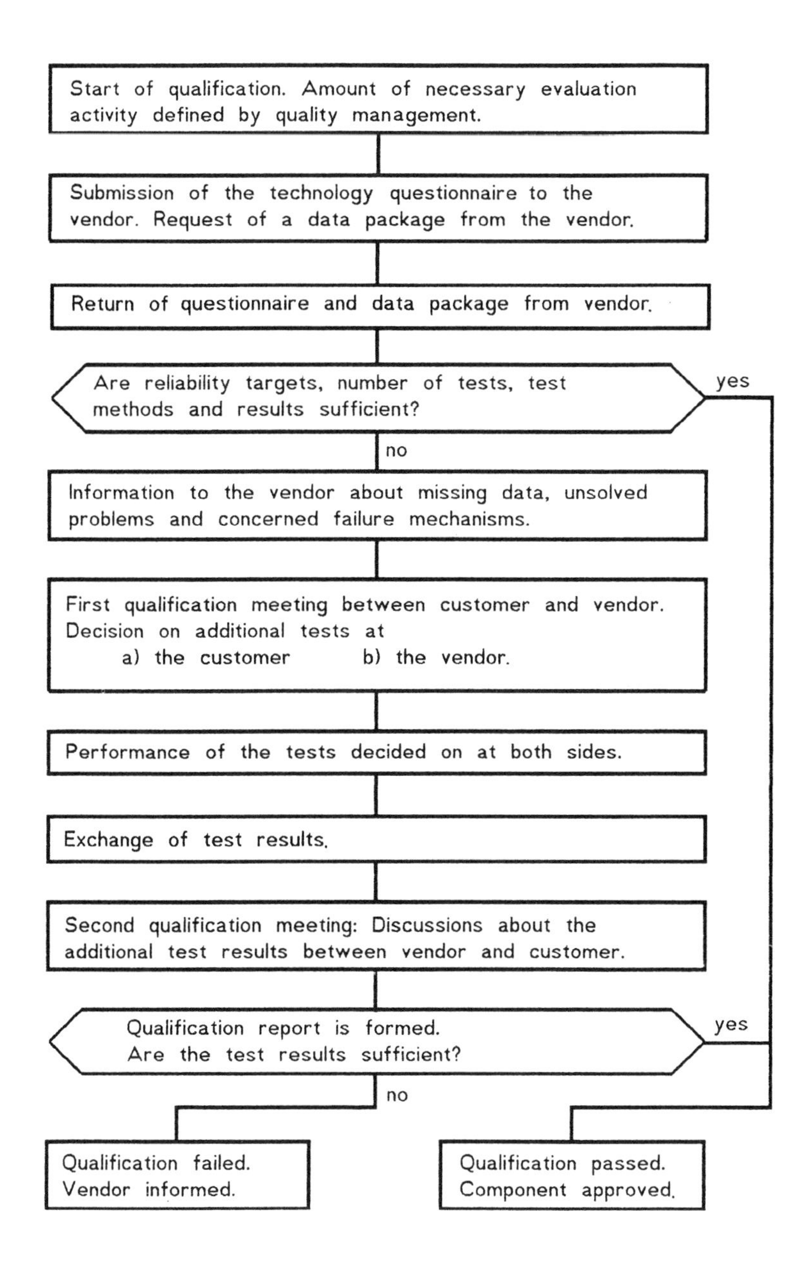

Figure 3.1 Joint qualification procedure.

Their activities, as helpful as they are, cannot completely replace the customer's own evaluation tests as presented in detail in the following chapters. The customer alone is responsible for the quality of products, but co-operation with the vendor or a test lab can reduce the amount of work considerably. Above all, a periodical requalification must be done by the vendor at customer's request. This has to be included in the specification, as shown in Chapter 8.

The strategy should be to maintain the ability and knowledge to do evaluation tests yourself but let the vendor do all routine work. You can only gain this knowledge, and keep it, by doing these tests at least once. Keep in mind the final goal of good quality of your products at low cost. This is by no means a trade-off. First comes good quality, then low cost. To make a good cake you should first look for excellent ingredients and then try to buy them cheaply. This may have not always been true in the past. For many people, low cost was the primary concern. But things change. Today, high quality is the main goal for each manufacturer.

The qualification usually consists of three parts:

- electrical and functional evaluation;
- test of packages to assure manufacturability; and
- environmental and reliability qualification.

It was the last item, the reliability qualification, where joint qualification began to become reality. This was because of the extent of the time and costs involved. The good experiences gained through this are encouraging both sides now to set up joint evaluation procedures for all three parts of characterization, as shown in Fig. 3.1.

The first step of this procedure is a very careful study and analysis of the vendor's data package. Ideally, all the parameters mentioned in the following chapters should be contained in a data package, but in reality it is usually less complete and contains only that data which the vendor's engineers consider important. A common agreement on the minimal requirements for a data package is useful to both parties.

Table 3.2 shows a rough scheme of the structure and the contents of a data package. The following chapters explain in detail how all parameters can be evaluated by the vendor or by the customer or to which critical points the customer has to direct attention when examining the vendor's data. The goal is to enable the customer to evaluate a component thoroughly and to feel confident enough not to expect any pitfalls when using it.

The first part of the evaluation package consists mostly of the functional and electrical parameters. They are split into two sections: First, there are design characteristics (laboratory data). These are curves of functions between static or dynamic parameters, usually drawn by the vendor's design department from a few samples. Some curves have to be drawn at different scales to show details of interest (Chapter 4). Then come product characteristics (ATE data) made by the quality department at several ambient temperatures. A larger sample size can be used when doing these tests, at least for standard circuits. So distributions of the measured values can be made by the vendor.

The next part of the data package consists of the reliability and environmental data (Chapter 5). A questionnaire given in Appendix D (page 272) may help the customer to discuss this item with the vendor. Section 5.9 describes some tests to be performed by the customer to check the feasability of the component to be used in the customer's production process. In the course of these chapters, some examples of lab data and ATE data are given.

One very important and cost-saving contribution to a joint qualification is a software database to be submitted to the customer by the vendor. This should comprise all parameters which are usually contained in the datasheet, the spice parameters of the circuit if applicable and all simulation data and circuit models necessary for the customer to perform a complete simulation of the products. As Fig. 2.2 illustrates, a simulation will speed up the development and improve the quality of the design considerably. It must be an objective of the negotiations between vendor and customer to arrange a compatibility of software that allows the importing of these data directly into the customer's software tools.

3.2 EVALUATION SURVEY AND INFORMATION SYSTEM

Normally an electronic device is composed of a great variety of different components. An overview of the most important single components is given in Table 3.3. In the second column of Table 3.3 the number of different types is listed, for all the components, which are used company-wide by a computer manufacturer producing medium-sized mainframes and workstations.

Of course, such a list is different for each manufacturer but the example given may encourage the user to compile similar lists for his

Table 3.3 Type count of the main components of electronic devices

Component	*Number of types (Variants)**		*Sub-total*	
Passive components				
Resistors	69	(3220)		
Capacitors	74	(1510)		
Inductors, transformers, filters	227	(750)		
Crystals	10	(110)	380	(5590)
			9%	(55%)
Discrete semiconductors				
Diodes, rectifiers	230	(490)		
Transistors, thyristors, triacs	310	(440)		
Optoelectronic devices, sensors	147	(200)	687	(1130)
			17%	(11%)
ICs				
Standard logic and interface	1387			
Microprocessors and peripherals	560			
Asics (types) ≤10k gates	72			
≤100k gates	102			
≤500k gates	13			
Programmable logic	46	(135)		
Oscillators	27	(250)		
Delay lines	20	(90)		
Memories (all kinds)	390			
Linear circuits and hybrids	420		2997	(3419)
			74%	(34%)
Total:			**4064**	**(10139)**

*Components of the same technology, function and package but with different values are designated as variants here and set in brackets.

or her own production. In most cases, the number of types will be enormous; in the example given above it was more than 10 000 positions if variants are counted too. In addition, there are always some components in requalification after a redesign done by the vendor. Due to an increasing division of labour, more and more sub-units are bought from OEM vendors, and these have to be evaluated too. A great number of single activities has to be performed by different groups of qualification people for all these component types, and the results have to be combined to reach a final conclusion as to whether the part can be approved or has to be rejected.

It is absolutely necessary for a device manufacturer to have a potent and effective database, an evaluation survey and information system (ESIS), to keep track of the status of the evaluation of all these component types. It is less advisable for him to program such a database himself. It is preferable to have it installed by a competent software company, based on a commercial database and retrieval system.

The organization of this database, however, must not be left to an external software company. It has to be specified in detail by the customer. This is a very important task as it will determine the effectiveness of the evaluation work for years to come. Many requirements of this database are set up from outside the quality department. The engineers who plan the data organization should not only think of new electrical and technological parameters which may become important in the future, but also of new requests from the market, for instance customer audits, or new legal requirements.

Because nobody can look into the future, the database has to be organized so that it is flexible and modular; so that not only its input and output facilities but also its structure can be changed easily. Another important requirement is data security.

The database must be transparent to the management and to all personnel involved. The inputs into the database must be made directly by the individual qualification groups, so paperwork is avoided and the data are always up to date. An effective and clearly arranged presentation of all data is very important to the general acceptance of the database.

Although it seems attractive at the first glance, the author does not recommend intermixing this ESIS database with other similar databases in the manufacturing area (section 6.3). The danger is that this would result in software that is both inflexible and difficult to maintain. It is better to keep the evaluation quality data separate from manufacturing-related quality data. A controlled link from ESIS to these other databases is by far a better solution than a global database. A clearly defined interface should be provided within the concept of the database to make such connections at any time.

Another link which may reduce the effort in updating the database is a link to automatic testers which perform electrical tests. This is explained in section 4.4. However, the main content of the database is not the test results in detail – these are documented in separate reports – but the status of the evaluation process.

Not all evaluation tasks, as they are described in this chapter, have to be performed on all components. At the start of a design the quality manager has to decide which of the evaluation activities is necessary for each new component. This is a decision of high importance and should be taken by experienced engineers. They have to consider all the activities as they are enumerated in section 2.3, the quality goal aimed at for that component and the related cost. This does not mean a trade-off between quality and cost, but an optimization of expenditure to achieve the maximum quality over all components.

Some principles to be regarded are given below.

1. The first time use of a component in the company requires a complete and extensive evaluation. Less activity is required in the following situations.

   (a) Similar components from the same vendor passed the qualification recently. If components belong to the same technological family and have roughly the same complexity, then some special tests have to be performed on only a few representative samples. For the other components, only a reference to these samples need to be made.
   (b) The component is used in low volume in a less essential application.

2. Reuse of a formerly qualified component in a new application requires less activity. Often it will be sufficient to compare the former test results with the special needs of the new application and to perform a more rigid and protocolled end test of the first prototypes.
3. Requalifiction of a redesign after a design change of the component by the vendor requires at first a thorough analysis of the design change note given by the vendor. The reason for the change has to be studied. Only critical parameters which might have been affected by this change have to be retested.
4. Periodical requalification should be done by the vendor as specified in the quality specification (Chapter 8).

Such activities, as enumerated in section 2.4, could be:

- to provide evaluation samples and assign them to the responsible evaluation group;
- to analyse the vendor's data package and check whether additional electrical tests are necessary;
- to prepare or check the availability of a loadboard and a test programme for incoming inspection;

- to start a joint qualification procedure with the vendor for reliability tests;
- to give samples to the production people – so they can check handling and solderability;
- to do electrical evaluation measurements if necessary;
- to obtain libraries for simulation of components; and
- to start a protocolled production in case of a reuse or requalification.

Each single task should be incorporated in the above mentioned database ESIS along with the following attributes: responsibility, starting date, target date or time limit, date of completion, result and a completion mark (to be filled when the task is finally completed). Other component-related data like internal numbers may be added.

There are four outputs of this database:

1. Automatically generated **qualification reports** are sent to the vendor in the course of the qualification procedure if the qualification indicates a 'fail' or to the purchasing department if it indicates a 'pass' (Fig. 3.1). This report may also contain some kind of grading of the evaluation result to give more information on potential risks when using a component:

   pass: 1: good, no restrictions for use
   2: fair, use tolerable
   3: fair, application restrictions to be observed
   4: single supply, back-up design required
   open: 5: in qualification
   fail: 6: obsolete component, do not use in new designs
   7: discontinuation of delivery apprehended
   8: quality insufficient, use not recommended
   9: potentially dangerous, use prohibited

2. Automatically generated **lists** show for all components:

   - open tasks (evaluation not yet started)
   - pending tasks (not yet finished)
   - expired terms (target date < actual date).

3. **Statistics** may be prepared, such as:

   - number of evaluations completed in a given time period
   - number of passed or failed evaluations per vendor.

Evaluation						
type:		vendor:	function:		typclass:	
74LS5004		ABC	Power driver 32 bit		MSI Bip	
started:	target:	finished:	report no.	date:	responsible:	result:
01/01/92	03/30/93				T512	5
remark:	first use: telecom receiver					

Datasheet/base:					com-plete
ordered:	target:	received:	revision no.:	responsible:	
		available	databook 91/92,	T512	+
remark:	no data base available				

Datapack:					
ordered:	target:	received:	revision no.:	responsible:	
01/01/92	03/01/92	02/15 /92	no revision no.	T512	+
remark:	electrical, reliability data, no technology data				

Datapack analysis:						
started:	target:	finished:	report no.	date:	responsible:	
02/15 /92	04/01/92	03/15 /92	AXY23654	17/03/92	T511	-
remark:	electrical data insufficient, own tests required, reliability lacking					

Sample order:					
ordered:	target:	received:	date code:	responsible:	
		01/01/92	4591 20 pc.	T512	+
remark:	free samples supplied by vendor				

Function/static parameter tests:						
started:	target:	finished:	report no.	date:	responsible:	
05/15 /92	07/01/92	08/01/92	AXT3458	09/11/92	T513	+
remark:	noise imm., ground noise, spec.static params.					

Reliability Questionnaire					
ordered:	target:	received:		responsible:	
05/01/92	08/01/92			R433	
remark:	target date expired reclaimed at 08/15/92				

Availability:					
started:	target:	finished:		responsible:	
03/01/92	04/01/92	03/15 /92		S523	+
remark:	two second sources available				

Figure 3.2 Evaluation survey and information system ESIS. A database is an urgent requirement to keep track of the qualification status of all components. The figure shows part of an input mask as an example.

4. **Evaluation data** can be stored for eventual legal requests.

There are many different ways of organizing such a database, so only a very simple examplc of some input masks and output lists for an MSI circuit is shown in Fig.3.2. This shows the reader what such a database might look like, although experts will surely find much more sophisticated ways of programming an evaluation survey and information system.

It is important, however, that the contents of this database are reliable. This means that the database has to be updated regularly so that its contents will always be consistent with the qualification result. The best way to achieve this is not to rely on goodwill but to install a software barrier which prevents the usage of a component by manufacturing without the database being completed and updated.

4

Functional and electrical evaluation of digital ICs

The first task when evaluating a component is to test all electrical properties: logic function, static parameters, dynamic parameters and some special properties. Because of the relatively small number of parts available for evaluation, no distributions or statistics are calculated; these are done later on incoming inspection. The purpose of evaluation is to obtain knowledge of all relevant properties of a component, even if these properties are not specified in the vendor's data sheet. A thorough knowledge of all properties of all components is absolutely necessary to make a mature, faultless and safe design. Even if a design works without problems in a prototype, problems may occur later during production if there are any obscurities. So the results of evaluation tests are needed by the design engineers; and they are given by the evaluating engineers in form of application rules, or as recommendations.

Another purpose of evaluation is to find weak points in the design of the component itself. This helps the customer avoid using dubious components in a design and it helps the vendor to improve components constantly. Additionally, the thorough knowledge of the component will help the user in analysing faults in the production phase and in becoming a competent partner for the vendor.

A further important use of evaluation is a comparison of different vendors as potential second sources. The design must be tolerant to deviations of parameters between vendors.

Whereas the test procedures for reliability, environmental evaluation and mechanical characterization are common for most kinds of components, the electrical tests are dependent on component type. It would go too far to give detailed advice in this book on how to evaluate all the types listed in Table 3.3; this can be done only for some important component types like ICs, especially large-scale ICs and asics. However, hints are given in sections 4.8–4.10 on how to apply these methods to other components.

For clarity, in this book, digital ICs are divided into different groups in a way which is commonly used (Fig. 4.2). There are off-

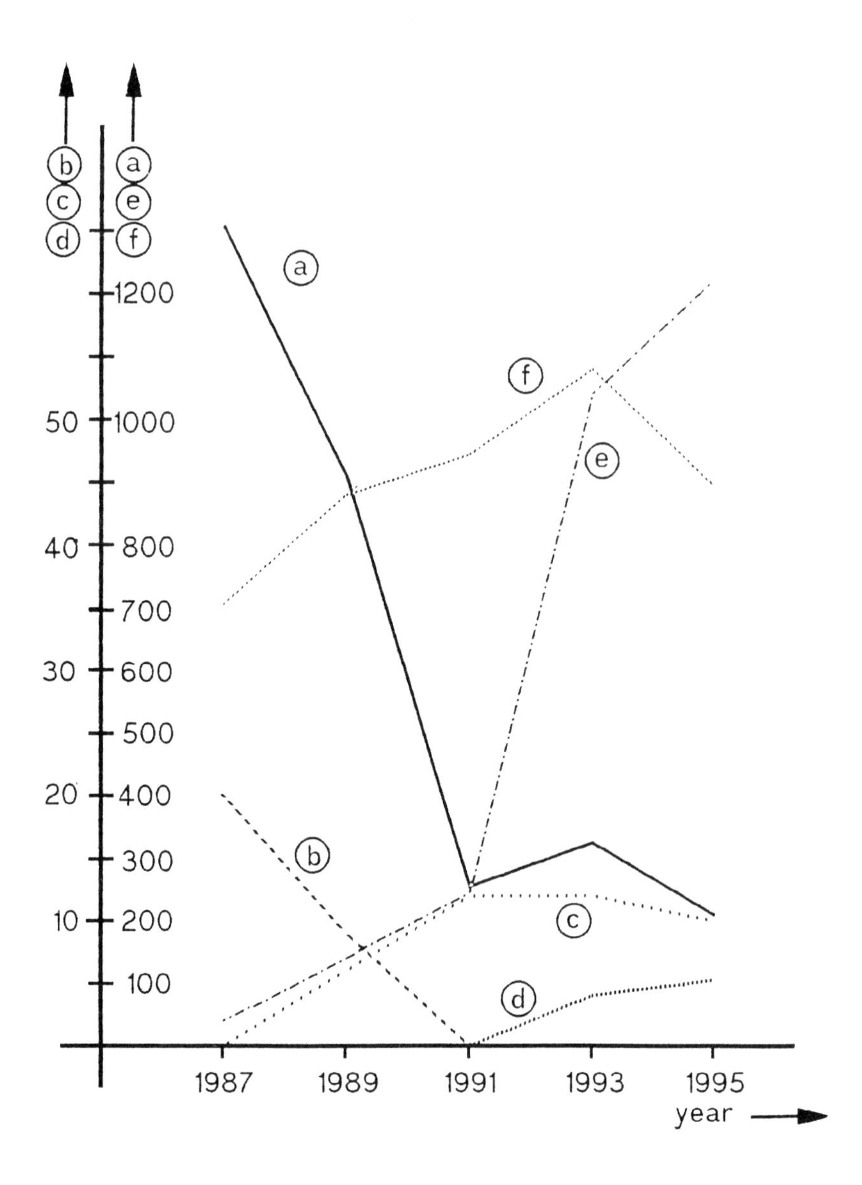

Figure 4.1 Advent of asics in modern designs: (a) number of standard logic components; (b) number of gate arrays with fewer than 2000 gates; (c) number of gate arrays with fewer than 20 000 gates; (d) number of gate arrays with fewer than 200 000 gates; (e) total number of available gates (the scale has to be multiplied by 1000); and (f) number of passive components.

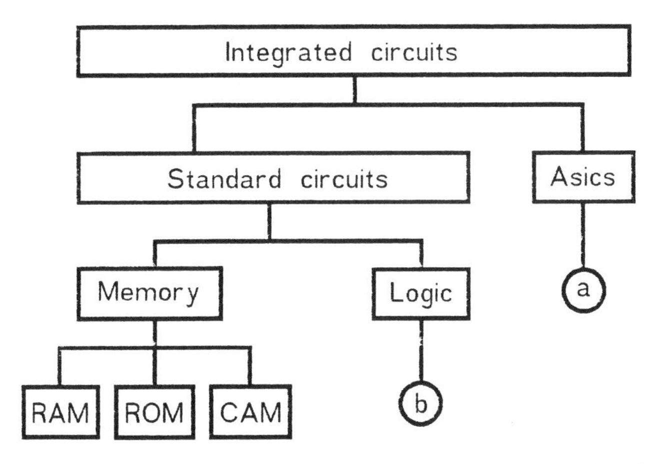

Figure 4.2 Classification of digital ICs – all are digital but electrical evaluation is different for each kind of component: for (*a*) see Fig. 4.30; and for (*b*) see Table 4.1.

the-shelf or standard circuits and application specific (or asic) circuits. Standard circuits are split into memories and logic circuits, the latter are subdivided due to their complexity as shown in Table 4.1.

Table 4.1 Classification of logic ICs by scale of integration

Abbrev	*Designation*	*Gate count*
SSI	Small scale integration	< 10
MSI	Medium scale integration	< 100
LSI	Large scale integration	< 1000
VLSI	Very large scale integration	< 10k
ULSI	Ultra large scale integration	< 100k
GLSI	Giga large scale integration	$< 10^6$

In the past, electronic devices consisted mostly of numerous SSI/MSI circuits. In modern designs, these parts of logic are combined in few asics. The advent of asics is demonstrated in Fig. 4.1, using an input-output processor of a mainframe as an example. It can be seen immediately that the standard components are reduced considerably in number (curve a) whereas the total number of available gate functions is increased (curve e). The number of asics and – not shown in Fig. 4.1 – the number of microprocessors and peripherals remained relatively constant or decreased slightly, but their gate count increased by a factor of a hundred. Gate arrays containg about 2 kgates (curve b) were replaced by asics with 20 kgates (curve c) and now gate arrays with 200 kgates or even more than 1 million gates are appearing (curve d). The number of passive components however, remained relatively constant within the same period of time (curve f). What is not shown in Fig. 4.1 is the reduction in size and the increase in performance of this device. The latter can be concluded from the increase of available gate functions (curve e).

Nowadays, microprocessors, asics and RAMs are the main components on printed boards, complemented by some MSI circuits as drivers for outgoing signals. Nevertheless, bipolar MSI circuits are used in the following sections to demonstrate the methods of evaluation. These circuits are rather simple and well known to engineers and did gain broad acceptance, and therefore are best suited for this purpose. The intention of the following sections is not to show methods of how to evaluate MSI circuits but to demonstrate some general principles of evaluation using MSI circuits as an example. The reader may easily adapt these methods to the kind of components to be used and either check the vendor's data or do some tests himself. The differences to MOS MSIs are mentioned here. In sections 4.6–10 the extensions for more complex circuits such as microprocessors and asics are given explicitely. These circuits are produced mostly in MOS technology. Some hints about how to evaluate discrete semiconductors and passive components are also given later, in section 4.10 (Hnatek, 1975; Mizko, 1986; Narud and Meyer, 1964).

4.1 LOGIC FUNCTION

The first questions which arise on evaluating an electronic component concern its logic function, because this is the purpose of the component. The quality engineer has to examine the vendor's datasheet to see whether the logic function is specified completely, exactly and understandably and for all conditions under which the circuit can be used. This is most important for complex LSI circuits, such as microprocessors, which will be discussed later, in section 4.6. The logic function is trivial for simple SSI circuits and for discrete semiconductors or even passive components, but with MSI circuits, like counters, shift registers, or adders, there could be some vagueness in the logic description. This starts with flipflops, where the outputs might be undefined when set and clear are active at the same time and leads to counters with the same problem for clear, load and clock inputs being active simultaneously. The quality engineer should ask the vendor for a written comment and/or make tests himself to clarify these functions; it is not sufficient to say that those input conditions are not allowed and therefore the function need not be specified. That may hold true for logic design engineers but the quality engineer has to set up a failure catalogue, as explained later in Chapter 6. Therefore he or she has to know about a possible malfunction caused by unallowed logic conditions.

One other aspect of quality is fault tolerance. An unintended wrong application of a circuit should not lead to an unknown result but to a well-defined error message, or better still the logic should be made insensitive to that error. The result of this evaluation of the logic function must be fixed in an addendum to the datasheet, accepted by the vendor, and supplied to the logic designer.

4.2 STATIC CHARACTERISTICS

Easiest to measure are static parameters, i.e. the response of the component to applied static voltages and currents. They are the basis of the electrical characterization of a circuit. Generally, these parameters are given by the vendor's datasheet and controlled on the vendor's outgoing inspection. Evaluation does not mean controlling these parameters again but trying to gain an insight into the behaviour of a component to obtain meaningful data with respect to quality. A good procedure is to measure electrical parameters as a function of other electrical parameters or other parameters like temperature, supply voltage or stress. So, in the sense of evaluation, parameters are more often a curve than discrete values. Practical examples of how to test electrical parameters and how to appraise the results are given in the following sections.

Static characteristics of ICs are taken mostly by a scope with curve tracer and print-out facility. But a simple and cheap test set-up, using an X/Y recorder, as shown in Fig. 4.3, can also be used. So it should be no problem for a customer to do some static tests himself. Whereas a recorder is purely static, a scope also includes transient currents to some extent. This may be important in cases of complex values of impedance. If there is any doubt, one can compare both results. Many component types have to be tested. Therefore it is advantageous and cost saving to use a universal flexible measurement set-up for DC tests. The test instruments (recorder, scope, loads, etc.) are connected by plugs and/or switches to the DUT pins. This is possible because signal distortion is not relevant for static signals. But nevertheless

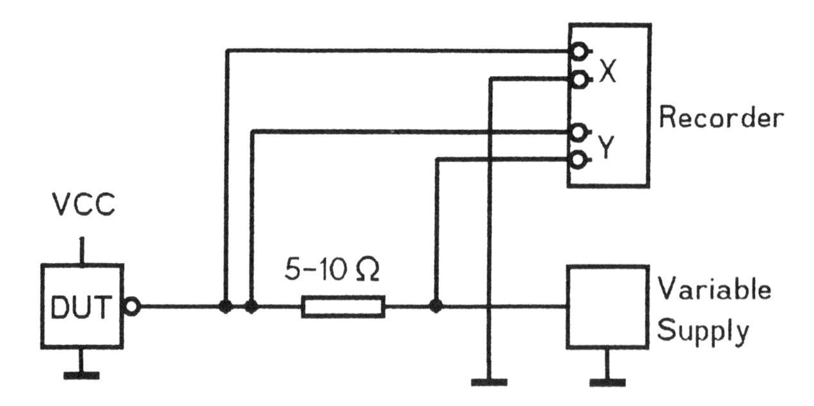

Figure 4.3 A simple and inexpensive test set-up to draw static characteristics (shown here for output characteristics).

special care has to be taken to design the measurement set-up. Even when taking static characteristics, oscillation can occur, which gives unnoted erroneous results. All pins have to be blocked in a suitable manner to ground. In critical cases, a scope should be used to ensure the absense of noise and ringing on all pins.

For components with typical diode characteristics, it might be necessary to draw several curves with different current scales, one scaled up to see small leakage currents and another scaled down to check the drive capability.

Warning: When taking static characteristics, the circuit under test may be driven into regions showing negative resistance. To avoid severe damage to the test object, it is very important and absolutely indispensable that the tester has current limitation properties.

Costs may often be saved by using an automatic tester, if available, to obtain static characteristics. This is dicussed in detail in section 4.4.

All the examples of characteristics given in the following subsections are not to compare specific components of specific vendors, but are generalized characteristics presented only to explain important considerations to be taken into account when examining measured static characteristics.

4.2.1 Input/output characteristics

The main static characteristics explained in the following subsections are input/output (I/O) characteristics, supply characteristics, transfer characteristics and some special static characteristics. They show significant properties for all types of active electronic components.

I/O characteristics are the representation of the internal impedance of an input or output. They are the current into or out of a pin as a function of the applied voltage. These characteristics are useful for the evaluation of integrated circuits in several regards.

First, they are the basis for the graphical determination of the waveforms on a transmission line connected to these pins. This so-called Bergeron method is a well-known practice and therefore will not be presented here; a short description is given in Appendix A. Less known is the use of a similar method to determine crosstalk which is also described in Appendix A. The knowledge of ringing and crosstalk, which can be derived from the characteristics, is of vital importance for deciding the layout of printed boards and it also gives the constraints for DA programs used for this purpose. The drive capability of an output determines the number of reflections and so is essential to calculate line delay (DeFalco, 1970; Singleton, 1968; Whittle *et al.*, 1992).

If an in-circuit tester is used for a board test, the tested devices or their drivers may be stressed beyond the datasheet limits. If, for instance, the tester forces a logic signal high to an input pin which is driven low from a preceding driving gate, then the driving gate might become overloaded. From the driver's output characteristics its overload can be guessed, which provides the input to a lifetime calculation. However, the manufacturers of in-circuit testers are going to reduce the likelihood of circuits becoming damaged by their testers.

Similarly, the possibility of connecting components of different technology, like bipolar and MOS, or of different voltage levels, like 5V and 3V circuits, directly without interface circuits can be estimated.

I/O characteristics also give an overview of the safety margin of typical values against datasheet limits. So you can obtain a feeling of how closely your vendor's design attained its simulated goals.

Second, they may indicate possible technologically weak points, as shown in Fig. 4.4. Every abberation from a normal curve – some are shown by dashed lines in the following figures – should be carefully discussed with the vendor, who has to explain the physical reasons for it and to prove its irrelevance.

Figure 4.4 shows an example of an input characteristics of a low-power Schottky circuit with hints on the meaningful points. Note that the scale for positive current is different from a negative one. Points a, b and c are datasheet values I_{IL}, I_{IH}, I_{IN}; point d V_{IC} defines the

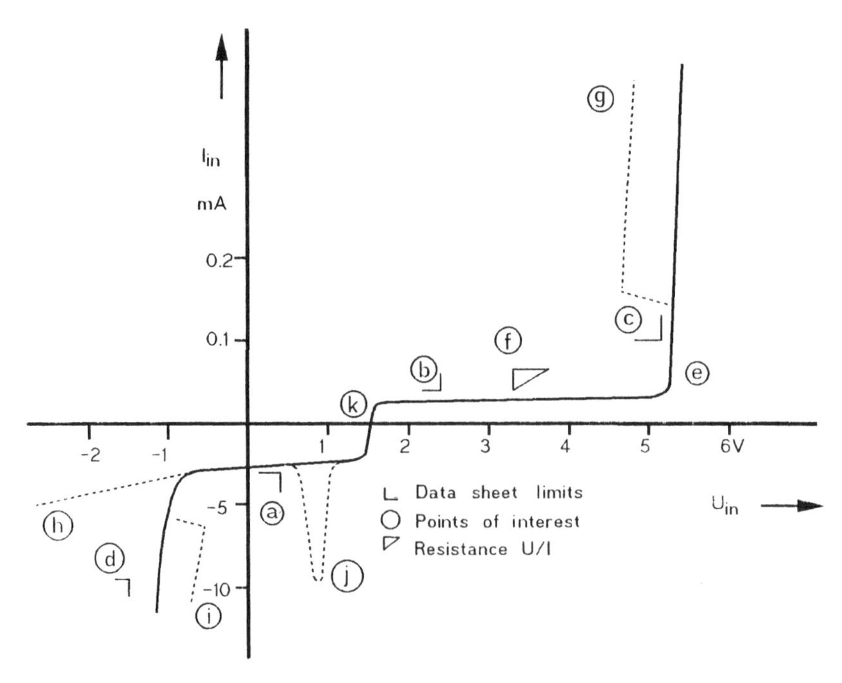

Figure 4.4 Input characteristics: input current as a function of input voltage showing some potential idiosyncrasies.

presence and quality of a clamping diode. The measured characteristics must be well within these limits. Points *e* and *g* show the breakthrough voltage (preferably greater than 6.5 V) and breakthrough resistance (preferably no negative regions). At point *f*, the input high resistance $\Delta U/\Delta I$ can be derived, which indicates poor technology when too low. Of some interest is the input open circuit voltage at point *k*. Dashed lines give examples of possible idiosyncrasies, which have been observed in several different components of several vendors but which are concentrated here into one figure for explanation. Such abnormal behaviour is caused in most cases by parasitic elements which do not show up in circuit or transistor diagrams. Point *h* shows the effect of a missing or badly clamping diode, point *i* a kind of latch-up and point *j* a current spike. These examples represent three categories of problems with different consequences:

- The absence of the clamping diode, whether caused by a bad design or by a production fault, is not tolerable for a customer.

Because it raises the danger of ringing, it will reduce the performance of the devices, even if the quality of the component is not affected.

- The latch-up effect shown at point *i* was caused by parasitic elements at one pin and for certain charges only. It was unknown to the vendor and not covered by his test program. The vendor committed himself to screen his components at first and then make a design change to solve the problem.
- Discussions with the vendor established that the current spike shown at point *j* was inherent to the design and has no adverse effect either on quality or on the applicability of the component.

The consequences for these categories should be:

- the use of the component is not recommended if the vendor is not willing to improve known drawbacks;
- the use of the component is recommended after the vendor commits itself to make some changes; and
- the use of the component is recommended without restrictions.

So, in some cases, the customer could help the vendor to find weak points in its design. In other cases, the complaint is a false alarm which causes unnecessary discussions. The vendor can help to avoid these if it provides more information about the details of its design to the customer together with the evaluation samples.

Two remarks can be made in this context. When measuring a so-called multi-emitter input of a TTL device, the logic state of the others, the so-called complementary inputs, has to be drawn to high potential. Also the input current of clock inputs may show irregularities caused by internal switching.

The idiosyncrasies shown in this subsection are examples only and the procedures described here are valid for other subsections too. The general principle is important: make measurements of all parameters and their dependencies; ensure you understand every detail of the result; if not, contact your vendor's design engineer. This general principle is valid for all kind of components.

Figure 4.5 shows the similar characteristics for the output high and low. Points *a*, *b*, *c* refer to the datasheet values I_{OL}, I_{OH}, I_{OS}. Point *d* is the output open circuit voltage. Point *e*, the maximum output current low, defines the driving potential of the output. Outputs with high driving capability do not show this kind of current limitation, but when outputs with a low driving potential have to drive dynamic loads

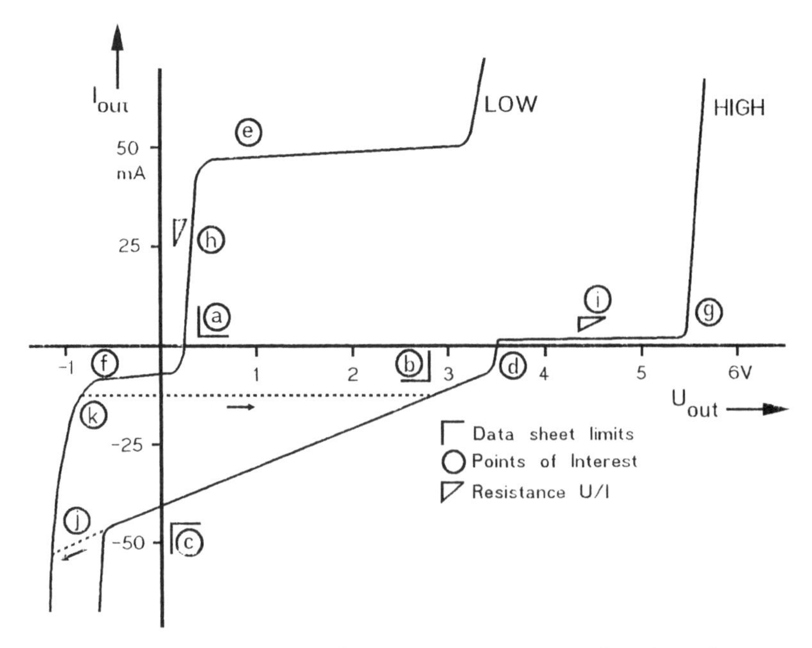

Figure 4.5 Output characteristics: output current as a function of output voltage for logic state high and low at the output.

this parameter is even more important than I_{OL}. Point *f*, the current at negative output voltage, is essential for the reflection of crosstalk at the output (Appendix A). Point *g* shows the output breakthrough voltage, points *h* and *i* the output resistance low and high, point *j* a potential hysteresis which has no adverse effect. Point *k*, which hints at a potential latch-up effect, is the so-called recovery voltage. It is an important parameter to estimate the additional delay which occurs under certain conditions when address bus lines are driven by bipolar drivers. (This is explained in subsection 4.3.5.) To measure this point you have to switch the output to a logical low state and apply a large negative voltage to reach point *j*. Then you have to change the inputs for a logical output high state. But in a critical circuit, the output remains low until the applied voltage rises to the so-called recovery voltage at point *k*. At that point, the output switches momentarily to the high state. The more negative this voltage is, the more delay can be expected.

An unintentional change of state during the testing of flipflop outputs has to be prevented by proper signals on preset and clear inputs.

If the output is in the tristate condition, then it should show high impedance for applied output voltages between −0.5 V to +3.5 V at least. High impedance between −0.5 V and +5.5 V would be even better.

MOS circuits are usually less burdened with problems than bipolar ones.

4.2.2 Transfer characteristics

The amplification or transfer characteristics are significant not only for integrated circuits but for all active four-terminal networks too. They describe the measured output voltage as a function of the applied input voltage: $V_o = f(V_i)$ as shown in Fig. 4.6 for an inverter as an example. Figure 4.7 shows the transfer characteristic for a non-inverting circuit as a second example. This measurement has to be done with loaded and unloaded outputs.

The primary reason for this measurement is the static noise immunity which can be derived from these characteristics. Static noise immunity is one of the basic parameters used in assuring the inherent quality of ICs. Sufficient noise immunity of all components is one precondition for the design of high quality devices. This parameter is necessarily controlled by the vendor when designing new components and also during the subsequent evaluation. Therefore, it should not be difficult to obtain these data from the vendor. But note that there are different ways of defining noise immunity exactly. A practical method is to define static noise immunity as the difference between the static low or high level voltage as defined in the vendor's datasheet and the so-called threshold voltage at point *b* in Fig. 4.6. The latter is defined as the voltage that has to be applied at the input, so that the output voltage equals the input voltage. This is shown graphically as the intersection of the measured transfer characteristic and the dashed straight line $V_o = (V_i)$ (point *a* in Fig. 4.6). A more mathematical definition is the point where the second derivative of the transfer function changes sign. Noise immunity plays an important role in ECL technology, where it is most critical (section 4.9). Here a chain of gates is used to define the so-called 'chain noise immunity'. It is far more important, however, to test noise immunity for different supply voltages and, for extreme temperatures, to obtain the worst case values (see points *c* and *d*). The influence of loads has to be considered too. For optimal noise immunity, the transfer characteristics should be steep, and the threshold voltage in the steepest part of the curve and in

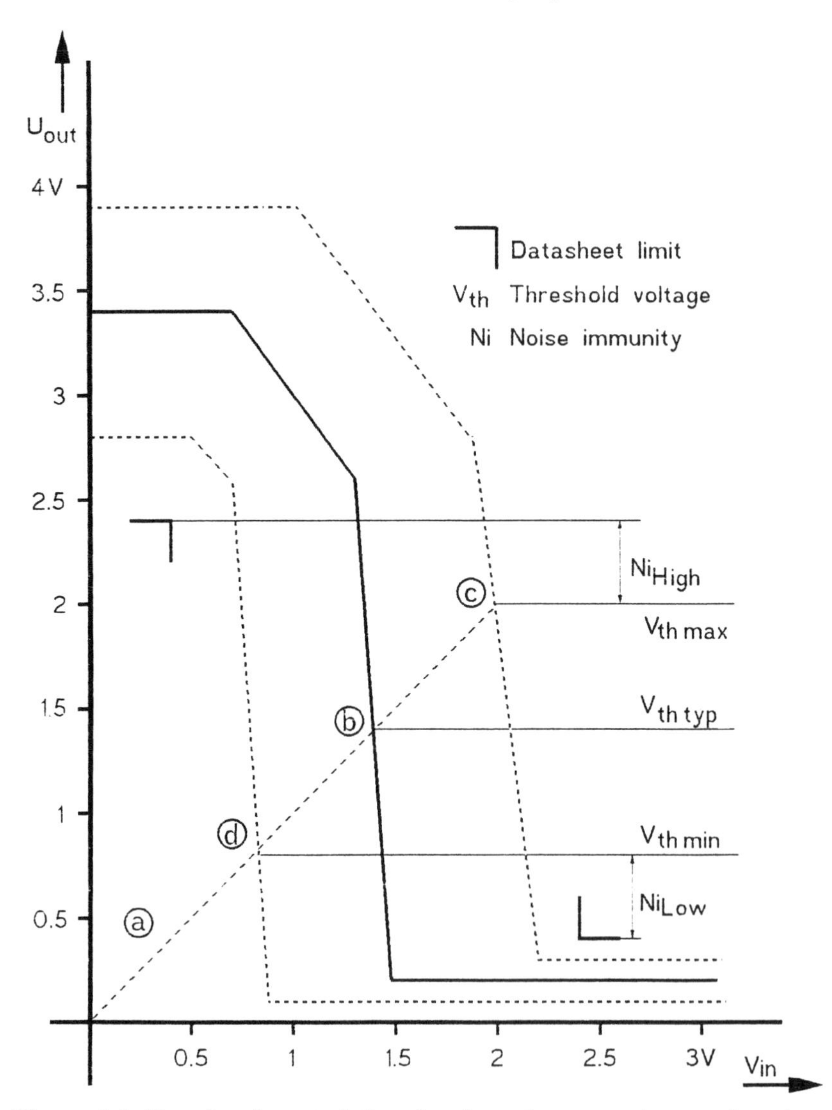

Figure 4.6 Transfer characteristics of an inverting gate – how static noise immunity can be derived from the measured threshold voltage.

the middle betwen high and low level. Some circuits show a hysteresis in the transfer characteristics (point *e* in Fig. 4.7). Hysteresis prevents oscillations when signals with slowly rising edges are applied to the input and it is sometimes used to increase noise immunity (Luecke, 1964).

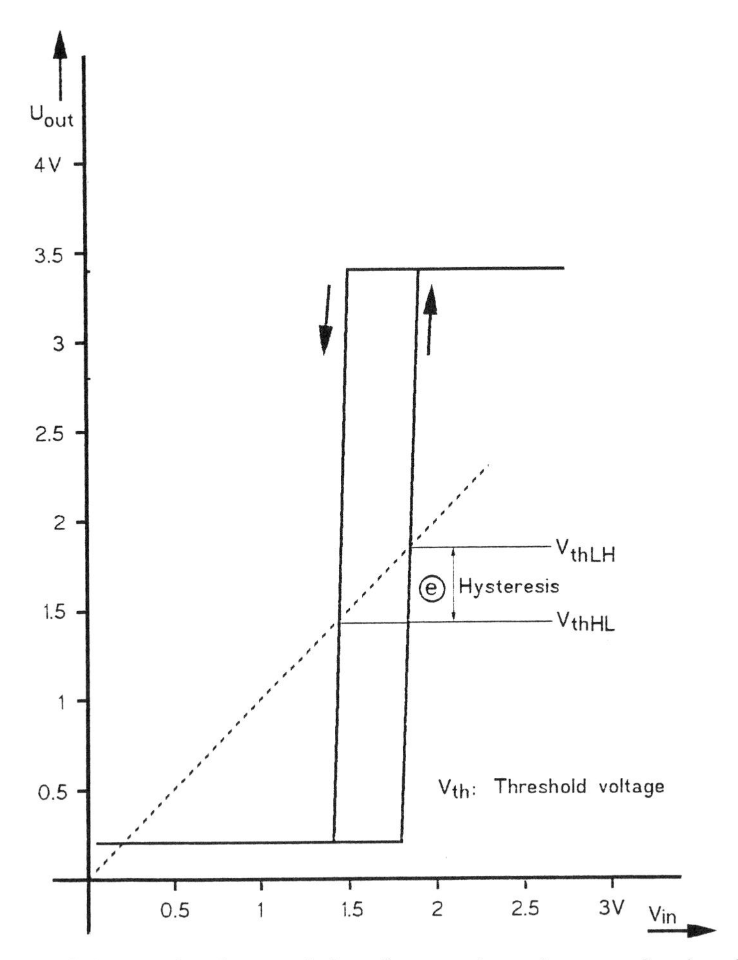

Figure 4.7 Transfer characteristics of an non-inverting gate. Scmitt-trigger behaviour shows different thresholds for rising and falling edges.

The transfer characteristics are of great use for designers who want to combine circuits of different technologies or supply voltages into one electronic system. The restrictions for driving conventional circuits from the newly announced circuits with 3 V supply voltage, for instance, can be derived from this curve.

When designing the test fixture, you always have to be careful to prevent oscillations which may give wrong results. This is critical for measurement of the threshold voltage because the circuit is driven into its active region.

Note that transfer characteristics from circuits containing sequential logic are less meaningful.

4.2.3 Supply current characteristics

The supply characteristics represents the current flowing from the power supply into the circuit as a function of the applied supply voltage. It may depend on the logic state of the circuit; SSI/MSI circuits have to be tested for high, low and tristate output. It should be a steadily rising function with a breakthrough well above the maximum rating of the supply voltage (point *b* in Fig. 4.8) and with no regions of negative resistance after breakthrough. Any peak in this steady function (point *c* of Fig. 4.8) is an indication of a transverse current. The reason may be conducting diodes at reduced supply voltage. This peak current should always be less than the nominal datasheet value of the supply current, otherwise it may cause an overload of the power supply during power-on ramping. This must taken into account for small devices containing many similar circuits.

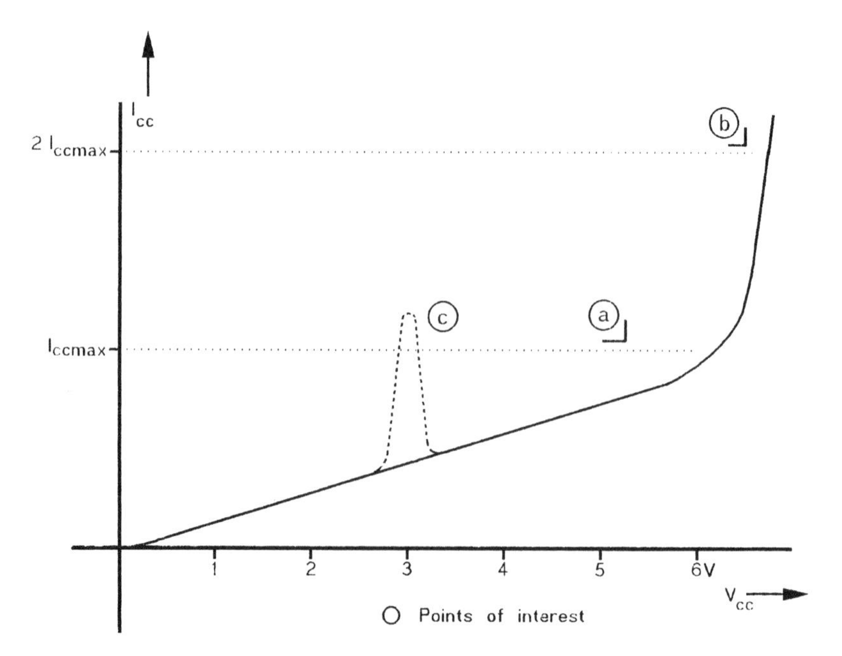

Figure 4.8 Supply current characteristics of a bipolar gate. The static supply current for components in MOS technology should be near zero.

Special attention has to be paid to supply characteristics on circuits with more than one supply terminal. The supply current has to be measured not only in normal operation but also with one of the supply terminals open or shorted to ground. If the values are inadmissibly high, then a certain sequence of power ramping or a tracking of supply voltages may be necessary.

MOS circuits should show an extremely small static supply current for all logic states. Any considerable static supply current points to a faulty transistor or to an illegal logic state. This can be used as a method to test MOS circuits by applying a function pattern to the inputs of the circuit in such a way that all MOS transistors are at least once in a low and in a high state, and by watching for an increase of supply current. While this method is good for analysis, it seems to be too time-consuming for incoming inspection because the large transient supply current during switching has to settle before the small steady-state supply current can be measured. But some automatic testers are available which overcome this problem by special provisions in hardware. They monitor the supply current continually while applying test vectors (Richardson and Dorey, 1992; Romanchik, 1993; Schiessler *et al.*, 1991, Soden *et al.*, 1992).

Recently, interest in this seems to have increased. This is because it may detect those physical defects which escape normal logic tests. Proposals have been made to integrate special current sensors on chip, or to minimize the number of test vectors, in order to speed up the test.

Note that the supply currrent may be less at elevated temperature. The rise in temperature caused by power consumption is a negative feedback to the supply current in this case. But a high temperature will increase the danger of technological defects like purple plague. It should be noted further, that the supply current is designated by the symbol ICC throughout this book for both bipolar and MOS circuits. In the literature, the symbol IDD is often used for MOS circuits.

4.2.4 Special static characteristics

The characteristics listed in the preceding sections are the main static characteristics which are significant for all kinds of ICs and also for all active four-terminal networks. They are repeatedly measured by the vendor during the internal characterization procedure before the circuits are released to production. So the user has a good chance of obtaining relevant and detailed data from the vendor within the evaluation data package and may restrict efforts to testing samples only.

But there are many other static characteristics which can give important application hints to the designers of electronic devices and be of great use during the debugging phase. These are often neglected and disregarded by the vendor. The customer has to make these measurements unless special agreements are made with the vendor. No general recommendations can be given here, because these parameters depend strongly on the kind of component and on the technology. Examples are given in the following subsections for TTL-like components.

I/O characteristics without power or during power ramping

This parameter is most important for tristate outputs. Many electronic devices have a modular structure, i.e. they are built up from different modules which are connected by a bus, mostly in the back plane. Often there is a request that modules have the ability to be plugged in or switched off while other modules are active. In addition, it is sometimes required that the data on the bus must not be disturbed by those actions. In such a case, the logic designer has to ensure a persistent tristate condition of all outputs. No spurious signals must be generated when modules are switched on or off. In order to guarantee this, the output characteristic has to show a high impedance in the full voltage range from –1 V to +5.5 V with supply pins open or grounded. Fig. 4.9(a) is an example for such a characteristic, with point 1 indicating a good sample, point 2 a rather bad one. This has to be included in the specifications to be negotiated with the vendors. Sometimes a designer expects a definite behaviour of his design during power-up. But an output which is low at nominal supply voltage may become high during power-on ramping as shown in Fig. 4.9(b) at point 3. Such a behaviour should be included into the design rules and has to be considered by the designer, e.g. when designing power-on reset circuits. A peak voltage of less than 0.8 V is tolerable.

Severe problems may arise when an electronic device consists of several modules which are connected to different power supplies and one supply fails. Again there is no danger if the outputs of this module show high impedance in the power-off case.

The information gained from the above characteristics is helpful in deciding whether special measures like protruding supply contacts for the modules are necessary.

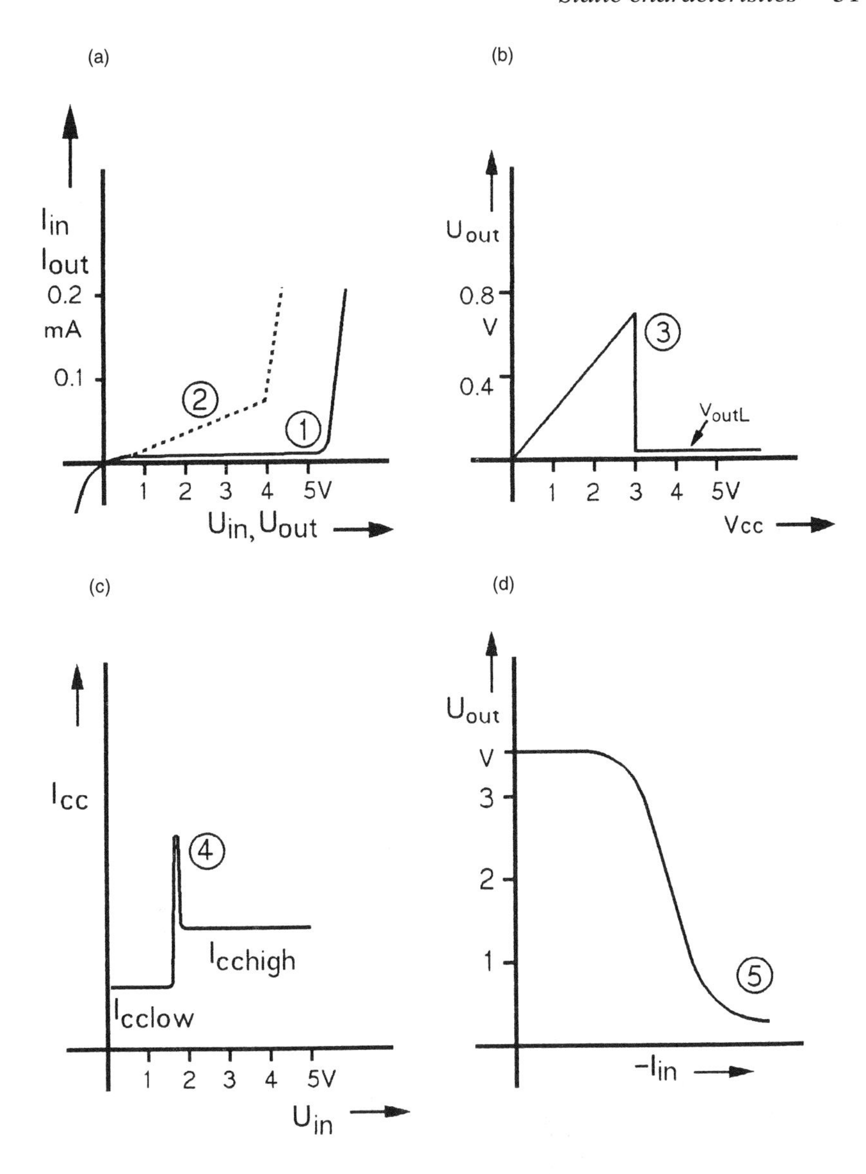

Figure 4.9 Special static parameters: (a) input and output current for supply pin open or grounded; (b) output low voltage while supply voltage is ramping up (loaded with 1 kΩ to V_{CC}); (c) supply current at threshold; and (d) output voltage for negative input current.

Supply current as function of the input voltage

Generally the output stage of a digital circuit consists of two transistors which are normally active in turn. During switching, it is unavoidable, in the case of high-speed circuits, that for a short period of time both transistors are conducting producing a supply current spike. This spike is the reason for providing filter capacitors on the boards. As will be shown in section 4.5.3, it is necessary to make dynamic measurements of these supply spikes. The above static characteristic gives a first indication of the possible amplitude of these spikes (Fig. 4.9(c), point 4).

Output low voltage for negative input current

Some ICs, above all those of low-power Schottky type, have a reduced output high state when negative current is drawn out of the input. This feature has to be controlled to prevent propagation of noise through several logic stages under adverse conditions (Fig. 4.9(d), point 5).

4.3 DYNAMIC CHARACTERISTICS

4.3.1 General considerations

Dynamic characteristics are the most important parameters, next to logic function and static characteristics, for the application of active electronic components. High speed is a deciding requirement for the customer and a leading sales argument for the producer of electronic devices. However, it is not the purpose of evaluation to make measurements of the dynamic datasheet parameters. This has to be done by the vendor and the measured data has to be included in the data package which is part of a joint evaluation, together with distributions of this data and detailed schematics. The evaluation of an electronic component should enable the quality engineer to look behind the parameters and understand how they are accomplished. He or she has to deduce the dynamic behaviour from the schematics and from the technological data – at least qualitatively. Operating the component in a test set-up and looking at the shape of input and output signals with an oscilloscope will help the engineer to achieve this goal. Again ICs shall serve as the examples in this chapter.

4.3.2 Test set-up

When measuring dynamic parameters by a benchtop test set-up, signal pulses from pulse generators are applied to the input pins of the DUT and load circuits to the outputs. Scope test heads are connected to both, inputs and outputs. High impedance active test heads are recommended when testing high-speed ICs. Again, it will be cost- and time-saving to use a universal test set-up with switchable or pluggable connections.

Figures 4.10(a) and (b) are a sketch and photo of such a test set-up. In this tower-like design, each socket pin is connected by a very short piece of wire to three connectors, which can be used to adapt a testhead, a terminated cable coming from a pulse generator and a load circuit. Enough free space is provided to put a small temperature chamber around the DUT. This set-up is sufficient for most high-speed circuits with rise times greater than 1 ns. Efforts have been made in the recently developed high-speed advanced CMOS circuits to slow down the signal edges. It can be assumed therefore that these universal test set-ups can be used for future circuits.

4.3.3 Delay parameters

The basic dynamic parameter given by the vendor is the delay time, the difference between the time when a signal is applied at the input of a component and the time when the corresponding signal appears at the output. This parameter forms part of the characterization data contained in the vendor's data package. Figure 4.11 gives an example from one vendor; other vendors supply similar data. The main problems of concern to the quality group are:

- correlation between vendor and customer, especially for high-speed circuits; and
- realistic delay observed in application against values specified in the datasheet.

The values given by the vendor are only meaningful if the test parameters are also specified in the datasheet. These are usually restricted to supply voltage, temperature and load circuit. This is not enough for high-speed circuits. The detailed layout of the test circuit and the load has to be supplied by the vendor. Sometimes it is necessary to exchange even the test set-ups between vendor and customer to come to comparable data. The parasitic capacitance and inductivity of different set-ups may cause differences in the results

(a)

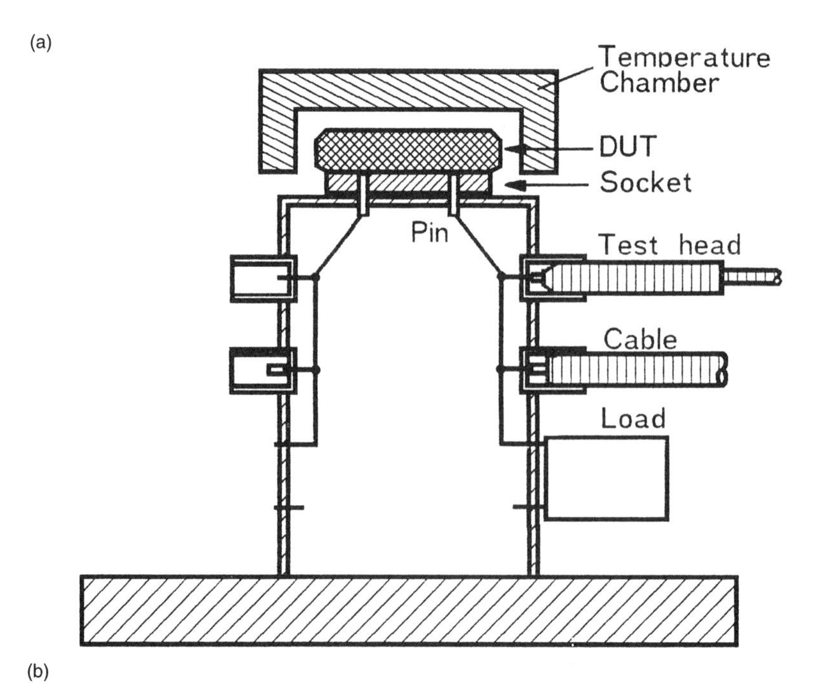

(b)

Figure 4.10 (a) Sketch and (b) photograph of a benchtop test fixture to test dynamic parameters of high-speed logic circuits.

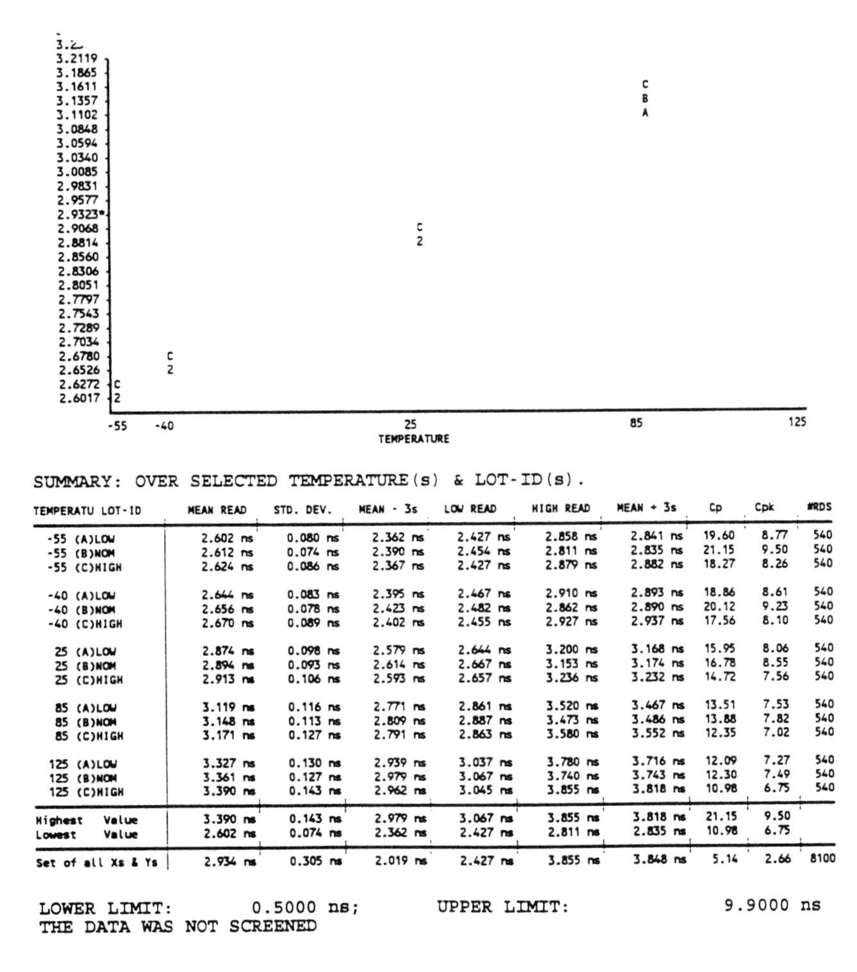

SUMMARY: OVER SELECTED TEMPERATURE(s) & LOT-ID(s).

TEMPERATU LOT-ID	MEAN READ	STD. DEV.	MEAN - 3s	LOW READ	HIGH READ	MEAN + 3s	Cp	Cpk	#RDS
-55 (A)LOW	2.602 ns	0.080 ns	2.362 ns	2.427 ns	2.858 ns	2.841 ns	19.60	8.77	540
-55 (B)NOM	2.612 ns	0.074 ns	2.390 ns	2.454 ns	2.811 ns	2.835 ns	21.15	9.50	540
-55 (C)HIGH	2.624 ns	0.086 ns	2.367 ns	2.427 ns	2.879 ns	2.882 ns	18.27	8.26	540
-40 (A)LOW	2.644 ns	0.083 ns	2.395 ns	2.467 ns	2.910 ns	2.893 ns	18.86	8.61	540
-40 (B)NOM	2.656 ns	0.078 ns	2.423 ns	2.482 ns	2.862 ns	2.890 ns	20.12	9.23	540
-40 (C)HIGH	2.670 ns	0.089 ns	2.402 ns	2.455 ns	2.927 ns	2.937 ns	17.56	8.10	540
25 (A)LOW	2.874 ns	0.098 ns	2.579 ns	2.644 ns	3.200 ns	3.168 ns	15.95	8.06	540
25 (B)NOM	2.894 ns	0.093 ns	2.614 ns	2.667 ns	3.153 ns	3.174 ns	16.78	8.55	540
25 (C)HIGH	2.913 ns	0.106 ns	2.593 ns	2.657 ns	3.236 ns	3.232 ns	14.72	7.56	540
85 (A)LOW	3.119 ns	0.116 ns	2.771 ns	2.861 ns	3.520 ns	3.467 ns	13.51	7.53	540
85 (B)NOM	3.148 ns	0.113 ns	2.809 ns	2.887 ns	3.473 ns	3.486 ns	13.88	7.82	540
85 (C)HIGH	3.171 ns	0.127 ns	2.791 ns	2.863 ns	3.580 ns	3.552 ns	12.35	7.02	540
125 (A)LOW	3.327 ns	0.130 ns	2.939 ns	3.037 ns	3.780 ns	3.716 ns	12.09	7.27	540
125 (B)NOM	3.361 ns	0.127 ns	2.979 ns	3.067 ns	3.740 ns	3.743 ns	12.30	7.49	540
125 (C)HIGH	3.390 ns	0.143 ns	2.962 ns	3.045 ns	3.855 ns	3.818 ns	10.98	6.75	540
Highest Value	3.390 ns	0.143 ns	2.979 ns	3.067 ns	3.855 ns	3.818 ns	21.15	9.50	
Lowest Value	2.602 ns	0.074 ns	2.362 ns	2.427 ns	2.811 ns	2.835 ns	10.98	6.75	
Set of all Xs & Ys	2.934 ns	0.305 ns	2.019 ns	2.427 ns	3.855 ns	3.848 ns	5.14	2.66	8100

LOWER LIMIT: 0.5000 ns; UPPER LIMIT: 9.9000 ns
THE DATA WAS NOT SCREENED

Figure 4.11 The distribution of delay values is a typical contribution to the vendor's data pack.

obtained. Of course, you have to observe all normally necessary precautions:

- Calibrate all instruments regularly.
- Check that the measured delay does not depend on pulse width or frequency.
- Load all outputs, even outputs not under test.
- Test all combinations of inputs and outputs.

One other often neglected test condition is the pattern sensitivity of the delay. Generally, the datasheet value is given for only one output switching. The number and position of other outputs switching simultaneously in the same package will influence the actual delay. The delay observed in application may be considerably more than that calculated from datasheet values due to this speed degradation caused by simultaneously switching outputs. This is most important for byte- or word-wide driver circuits and for asics as well.

Additional delay parameters, not specified by the vendor, but necessary to obtain the utmost performance, are:

1. reduced delay for lightly loaded outputs ($F_O=1$, $F_O=2$) – whether this can be measured directly depends on the capacitive load of the test fixture (otherwise it must be calculated by extrapolation using two measurements with heavy loads);
2. pulse width distortion ($t_{pLH}-t_{pHL}$) has to be known to define the minimum width a pulse can have without being extinguished;
3. delay differences between:

   (a) different circuits within one chip,
   (b) different outputs driven by one (enable or clock) input, and
   (c) different (complementary) inputs driving one output,

   are esssential to calculate the clock skew for the usual synchronous logic;
4. delay difference when unused inputs are connected in parallel or to a fixed potential – do not allow unused unconnected inputs;
5. delay difference for EXOR gates with the unswitched input connected to L or H potential; and
6. delay difference for multiplexer circuits when outputs are in phase to address (select) inputs or not.

All these delay differences have to be taken into consideration by the logic designers when making any exact delay calculation and the quality group has to provide this information to them. It has to be stressed here that all these topics apply to asics as well.

Special care has to be taken when components are used which are selected by the vendor to meet the required dynamic parameters. This may happen in advanced circuits, when the vendor failed to meet design goals or when a vendor offers versions of the same circuit but with better dynamic performance, e.g. clock frequency, at a higher price. As a consequence, the distribution of the selected parameter is not gaussian, but cut off. Figure 4.12 illustrates this situation for a critical delay time t_{pd}.

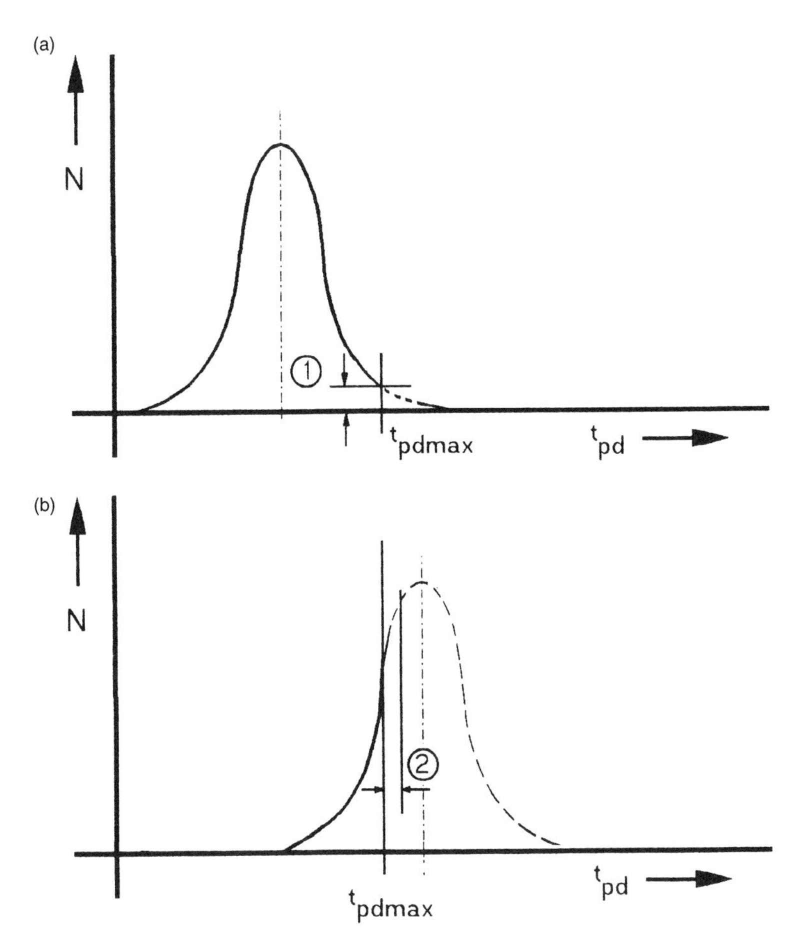

Figure 4.12 A cut-off parameter distribution (b), as a result of a selection by the vendor, makes ship-to-stock more difficult than a gaussian distribution (a).

Normally, as shown by curve (a), the distribution is well below the maximal value t_{pdmax} of the datasheet and the intersection point 1 reflects the guaranteed failure rate. In a cut-off distribution, curve (b), a small uncertainty in delay measurements will cause high differences in defect rates. Such a situation is detrimental to a ship-to-stock procedure and forces the customer to maintain incoming inspection (section 6.4.4). To avoid such difficulties, a safety margin to the datasheet value should be agreed upon (point 2 of Fig. 4.12). Chapter 8 gives advice on how to specify this.

Far away from practical application are the values given for enable and disable times of tristate outputs. Take the example given in Fig. 4.13(a).

Data is transmitted from an asic A through a bidirectional driver B, a line C, and second driver D to another asic E. Now the direction of data flow should reverse as soon as possible. Try to calculate, from datasheet values, at what time the enable and direction inputs of the four circuits has to be switched without outputs driving against each other, and without unnecessary floats.

Usually in the datasheet no delay time t_1 is specified from disable signal at input to high impedance at output without the time constant t_2 of the load as defined in Figs 4.13(c) and (d). For the case shown here, the result is very simple. You can switch EN_1, DIR_1 and DIR_2 at the same time, because the bidirectional drivers are all non-inverting and so the signal level is the same on all the interconnecting lines. Therefore the outputs will not drive against each other. Before enabling EN_2 you have to wait for the time t_1 of driver D. The signal delay from the output of asic E to the input of asic A is: delay t_3 of driver D plus line delay t_L plus delay t_3 of driver B. (Fig. 4.13(b)). The situation would be quite different – and worse – if the drivers were inverting the polarity of the signals. Bus contention will cause additional delay. This is one example of how the quality engineers have to consult the logic designers.

Dynamic parameters of sequential circuits like counters or shift registers are sometimes misleading for the user. In most cases, the maximum frequency of a counter is defined for a straightforward count without feedback. If the carry output is used to reload the counter, then the maximum frequency will be considerabily less. To exceed it, will lead to a miscount.

4.3.4 Transition time (rise and fall time)

A very important parameter for new high-speed circuits is the minimal rise and fall time at output. Very steep pulse edges will cause ringing and excessive noise on other adjacent lines. Special sophisticated circuit designs are made by the manufacturers of high-speed circuits to limit the steepness of the edges. Circuits with edges less than 1 ns for 3.5 V amplitude will be difficult to use reliably.

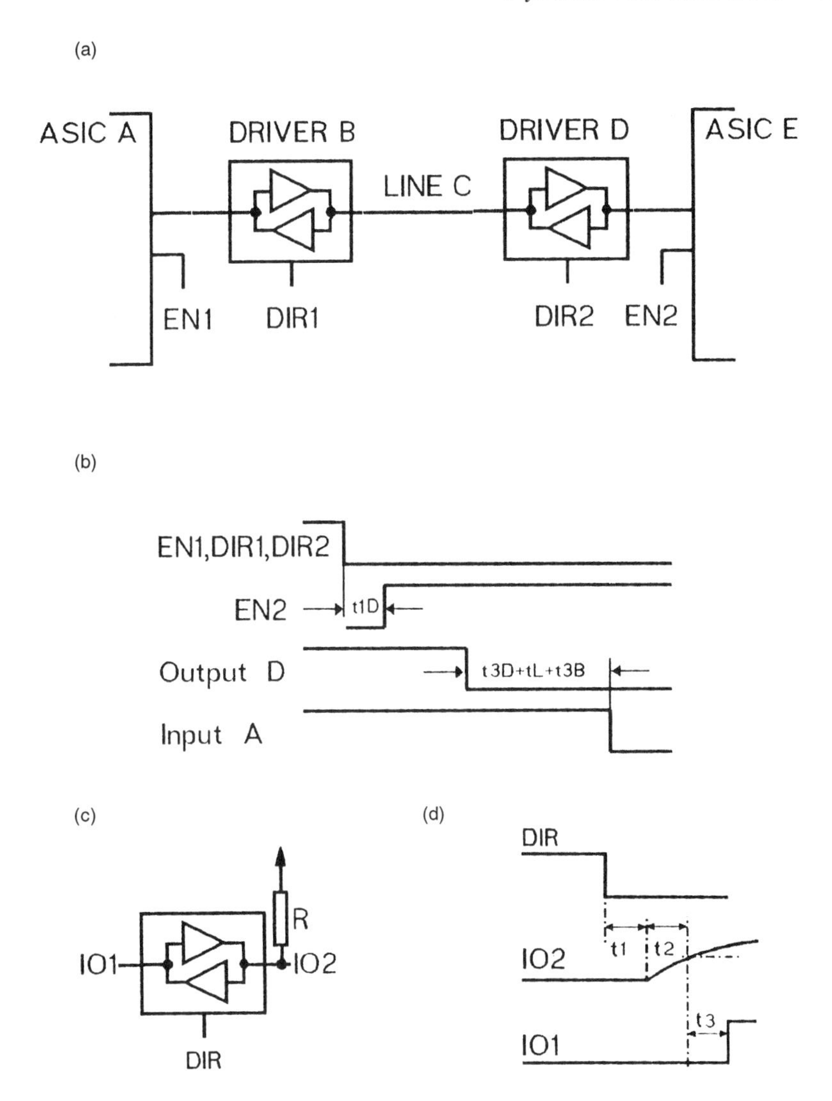

Figure 4.13 Bidirectional data transmission between asics. When the direction of data flow changes, the pulse diagram (b) shows the timing relations of the circuit (a) based on the delay parameters (d) of the bidirectional driver (c).

The maximum allowed transition time of an input signal is an important dynamic parameter too. Signals which are too flat will cause oscillations on gates and double triggering on flipflops.

Acceptable limits are:

- 1 μs/V for low-speed circuits
- 100 ns/V for high-speed circuits

So a rise time of 500 ns of the input signal is suitable and often used to test this parameter.

4.3.5 Logic spikes and noise may increase delay

Logic spikes are a well-known phenomenon to logic designers and are avoided by taking appropriate measures:

- multiplexers may have decoder spikes at address change; and
- counters may have decoder spikes at carry outputs.

The logic designers have to introduce some delay until they can use the signal. Because this is often disregarded, the quality engineer has to include this in the application rules. Some vendors claim to have spike-free circuits, but it is wise to check this on evaluation, especially for the case of unsymmetrical output loading.

When using high-speed circuits unexpected sporadic oscillations may occur. The reason is in-phase feedback. For example, the coupling from the output to a non-switching low input of an EXOR-like function may cause such oscillations. The phase shifting methods used by designers of high gain analog amplifiers will help here. In the above example, capacitive feedback by a small capacitor will terminate the oscillations.

Large negative-going noise pulses, when applied to a low-level output, may cause a considerable delay to a positive-going edge. Not all component types are affected to the same degree, some low-power outputs are more concerned than others. The noise pulses will drive the output voltage far into the negative branch of the output characteristics. In section 4.2.1, it was shown how the ability of a circuit to withstand this noise can be derived from the output characteristic. If the output voltage becomes more negative than the recovery voltage V_F, (Fig. 4.5), then this delay will be noticeable. If the affected output belongs to a counter, then a miscount may also occur.

An example for such a case is given on Fig. 4.14(a). The *n* address lines of a memory bank are driven by a driver or register with a total

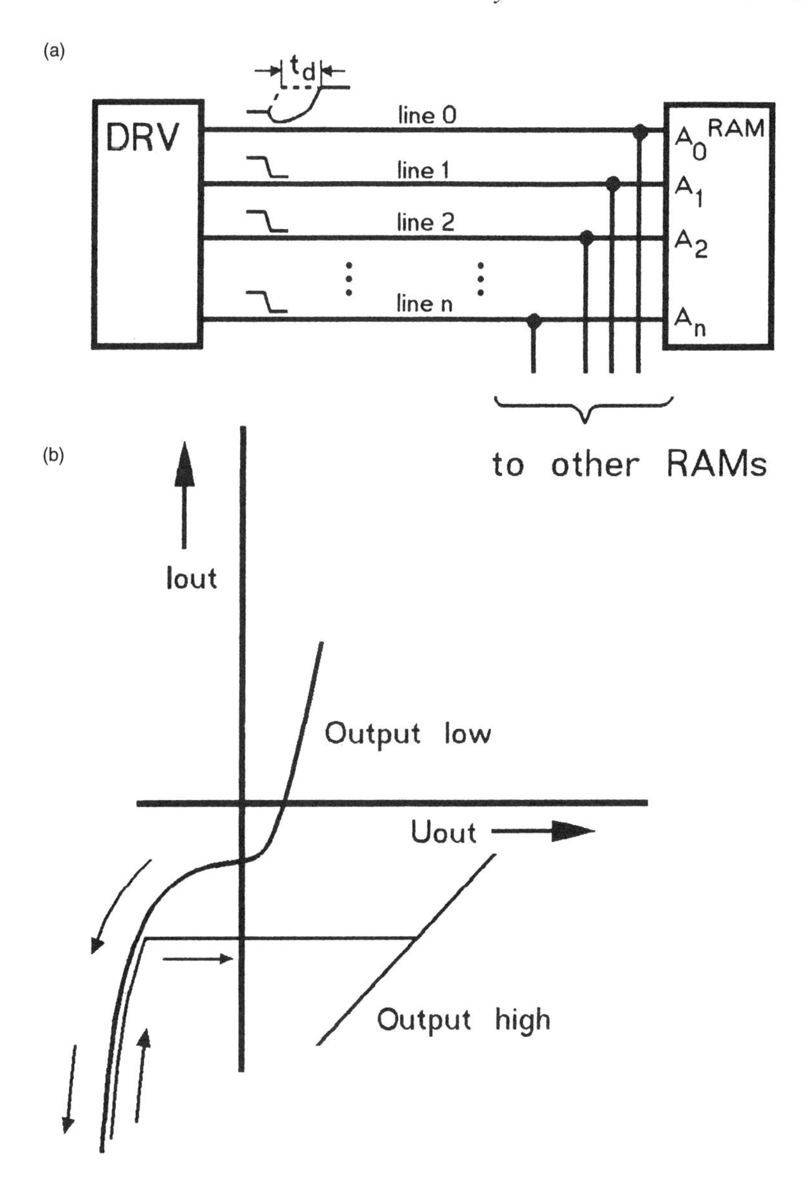

Figure 4.14 Increased delay caused by noise: (a) an example of how crosstalk may induce noise at the output of a driver; and (b) output characteristics of the driver enable estimation of whether an additional delay has to be considered.

line length of more than 50 cm. Assume that all addresses change their state at the same time, all lines but one switch from high to low state, one line from low to high. All the address lines are usually heavily loaded and relatively long. The low-going lines will cause a large negative noise pulse due to crosstalk on the line which should go high. But this noise pulse drives the output, which is still low, far into the negative region of its characteristics, as shown in Fig. 4.14(b). This will prevent the output from coming up and delays the positive edge by 1 to 20 ns.

This effect is well known to the vendors but not specified. The customer should check this on evaluation and avoid using such components. If a noise voltage of –0.5 V is tolerated, then the measured recovery voltage V_F will be 0.5–0.8 V for an acceptable circuit and 0.3–0.5 V for a bad one.

4.4 EVALUATION BY AUTOMATIC TEST SYSTEMS

4.4.1 The benefits of automated testing

All the measurements for evaluation of components, as described in the previous sections and also in the following chapters, are usually done either manually or semi-automatically: static parameters by a X/Y recorder or by an oscillocope, dynamic parameters by a benchtop system, for instance by a combination of pulse generators and oscilloscopes. The advantages are given below.

- Manual evaluation is very flexible and all kinds of components can be thoroughly tested and inspected.
- The evaluation of newly designed parts can be done at once without much preparation.
- The test engineer can attend immediately to any irregularity detected.
- The test and the analysis of the results can be done in one step by the same engineer, so a later repetition of measurements to clear up ambiguities is seldom necessary.

But there are disadvantages too.

- The manual evaluation of components necessitates skilled persons and is time consuming and therefore very costly, especially for components with high pin counts.
- All combinations of input and output pins have to be tested, yet it

is often only one combination which shows negative results. Therefore many very similar measurements have to be performed. The test engineers may become frustrated because most evaluations have positive results, so their extensive and laborious work attracts little attention. But any abatement of care will reduce the value of their work considerably.

A better solution is the utilization of an automatic tester (ATE) for all kinds of electrical evaluation. This is preferable if evaluations have to be done repeatedly or if many evaluations of the same kind are required. This happens at requalification of components after a design change or evaluation of different components belonging to the same family. ATE is useful also for doing electrical tests which are required at end point of life tests. Further, it is an excellent basis on which to share the work with the vendor, because it presents an unequivocal interface. An ATE is indispensable when evaluating complex components with a high pincount like LSI circuits or asics. It is of some advantage also in testing surface-mounted packages avoiding adapter problems. A tester which is used for incoming inspection and troubleshooting can be used for evaluations as well, so no extra investment is needed (Stover, 1984).

An ATE needs software. It is advisable to create this software by means of a program generator, because much flexibility is necessary for the low part count and the many part types, whereas short test time is not a main requirement. Such a program generator is usually available from the tester manufacturer but it can easily be written by a skilled test engineer. The test software should have a modular structure, then tests and test parameters can easily be added, deleted or changed. Appendix B gives an example for a modular structure.

It is not only the measurements themselves which should be performed automatically but all further steps of evaluation including the analysis of the results. This can be done by the tester itself, using its controlling computer, or the results can be written into a file and all further analysis done later by a separate (cheaper) computer.

The first step in this analysis is a graphical presentation of the results. The static and dynamic characteristics, which are measured point by point, are plotted. A two-dimensional plot, i.e. a set of curves with added values, may give a better overview, e.g. the dependency of delay on supply voltage and temperature, or another kind of presentation made possible by the graphical creative style.

This first step of automatization yields a considerable reduction in cost and time. No instruments have to be put in place, adjusted and operated; no paper has to be fed into a print-out facility. But all the plots and graphs have to be analysed manually by a skilled person. Therefore the second step is an automatic analysis of the measured results by the computer. Only the outcome of this analysis – pass or fail – has to be interpreted by the test engineer. There is a wide range of engineering skills for deriving programs for this task. Only a few suggestions can be given here. Take the supply characteristics as an example for the analysis of a static characterization.

1. A band gap can be defined to set an upper limit for the measurements by

$$I_{CC(meas)} < 0.9\, I_{CC(max)} * \frac{V_{CC(meas)}}{V_{CC(nom)}} + 0.1\, I_{CC(max)}$$

for $V_{CC} < V_{CC(nom)}$ and

$$I_{CC(meas)} < 0.9\, I_{CC(max)} * \frac{V_{CC(meas)}}{V_{CC(nom)}} + 0.6\, I_{CC(max)}$$

for

$$V_{CC(nom)} < V_{CC(meas)} < 6.5\,\text{V}$$

where $I_{CC(meas)}$ is the measured current at $V_{CC(meas)}$ and $I_{CC(max)}$ is the maximal value defined in the datasheet for $V_{CC(nom)}$. This upper limit is shown in Fig. 4.15(a).

2. The first and second derivatives can be calculated to detect inadmissable slopes or eventual spikes in the characteristics.

$$\frac{dI}{du} = \frac{I_{CC(n)} - I_{CC(n-1)}}{\Delta u}$$

where $I_{CC(n-1)}$ and $I_{CC(n)}$ are two successive measured values and Δu is the voltage step. The result of these calculations is plotted in Figs 4.15(b) and (c) in case of a current spike. The condition for a circuit to pass evaluation is:

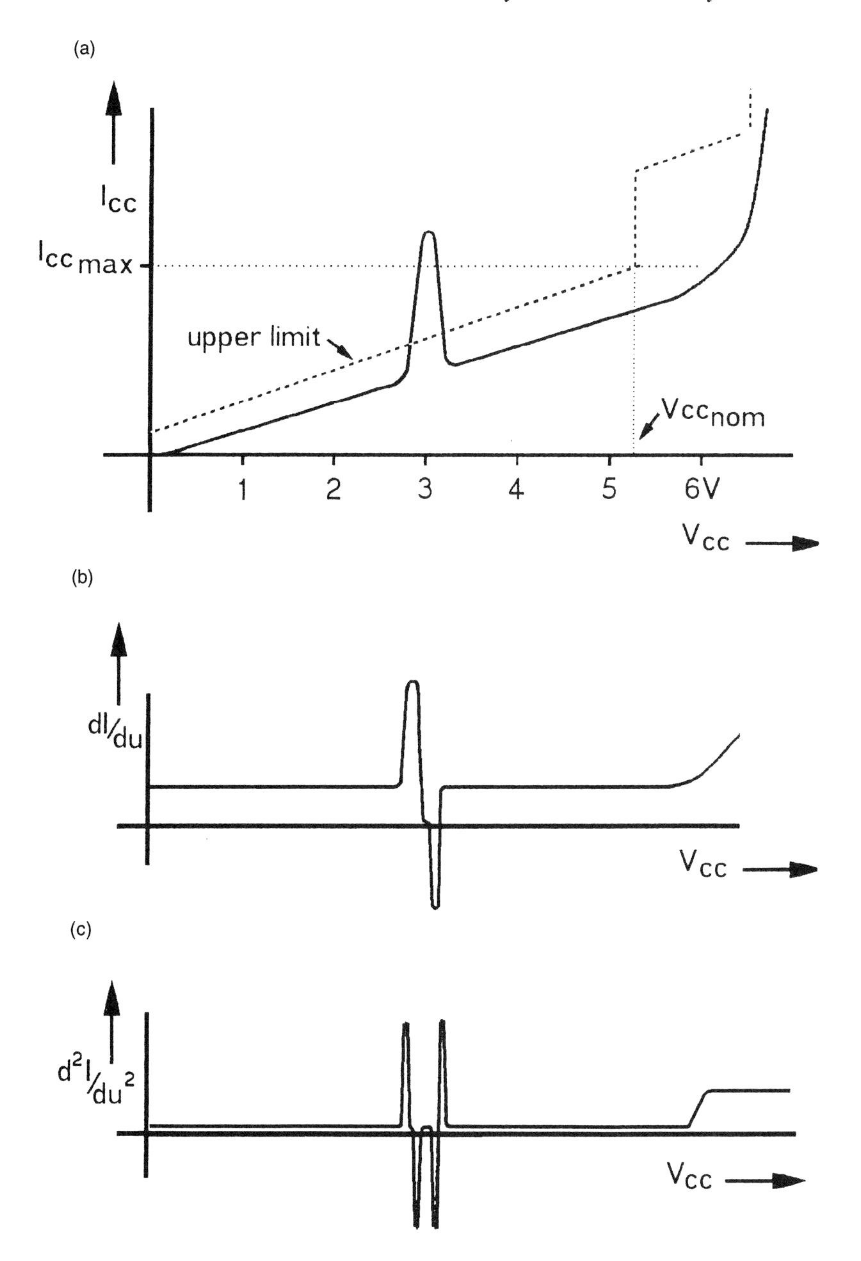

Figure 4.15 Automatic analysis of static characteristic – differentiating the measured curve reveals any discontinuities: (a) supply current characteristics; (b) first derivative; and (c) second derivative.

$$\left| \frac{d^2 I}{du^2} \right| < 0.1 \, I_{CC(max)}$$

Only these final results have to be printed and transferred and stored in the database ESIS.

The measurements for MSI/SSI ICs, described in the previous sections, were only examples to demonstrate how extensive and detailed evaluation tests give the quality engineers full understanding of all properties of the circuits and enable them to react to all application problems which may occur later. The same principles are valid for all other components such as LSI/VLSI ICs, analogue components, discrete semiconductors and even passive components.

The following calculation gives an impression of what cost reduction can be reached for SSI/MSI circuits and also for LSI/VLSI components (which are described later in sections 4.6 and 4.7).

4.4.2 Cost comparison

The time needed for an electrical evaluation of an IC type depends on many factors. The following cost comparison is based on a complete evaluation of electrical properties only. It is assumed that it is performed by a medium-skilled person using vendor's information and data package, as is now commonly available. Comparisons are made between manual tests with benchtop instruments (curve tracer, pulse generators, oscilloscopes, logic analyser) and tests using an ATE with effective evaluation test programs, as explained in section 4.4.1. A complete evaluation means performing all the tests described previously as far as they are necessary to verify and complete the vendor's data package, the analysis and valuation of the results and proper documentation. The following data come from companies who have been using ATE for evaluation for many years.

The time needed to evaluate a simple SSI circuit. like a driver gate, will be 2–4 days; for a more complex MSI circuit, like a counter or shift register, 4–12 days. This time increases by 75% if one or two second sources have to be evaluated later and by a further 25% for one redesign being requalified. The time can be considerably reduced if the components belong to a family of similar circuits, because some tests have to be performed on only a few members of this family. The time required to evaluate LSI/VLSI circuits, like microprocessors or peripherals, shows great variance. Some peripherals are very difficult to evaluate. Assume 20–60 days here.

If ATE systems are used, then the time for the test itself is negligible. To write and debug a sophisticated evaluation test program using a program generator needs 0.5–2 days for SSI and MSI circuits. For LSI/VLSI circuits about 20 days are needed, if the methods described in section 4.6 are used.

Table 4.2 Automatic test system saves evaluation time

	SSI	*MSI*	*LSI/VLSI*
Electrical qualification of one type	2d – 4d	4d – 12d	20d – 60d
Electrical qualification including second sources and redesigns	6d	15d	50d
Number of components qualified per year by one engineer	100	25	5
Time for one evaluation test program	0.5d – 2d		20d
Number of components qualified per year by one engineer using ATE	200		10

Table 4.2 lists in the upper half:

- the time to evaluate one component type manually (in man days);
- the mean time to evaluate one component plus two second sources plus one redesign in total (in man days); and
- the number of components belonging to a family which can be evaluated by one engineer or technician within one year (approximately 200 working days).

In the lower half Table 4.2 lists:

- the time to evaluate one component by an ATE system, which means the time to write and to debug the test program – to evaluate second sources and redesign requires virtually no extra time; and
- the number of components which can be evaluated per year, considering that the test programs for similar circuits can be derived by making minor changes.

If a mixture of 40 SSI, 30 MSI and 5 LSI circuit types are assumed as a typical mix to be qualified within one year, then it can be derived from Table 4.2 that about three engineers are needed in case of benchtop testing and only one engineer with ATE. Saving on qualification costs does not mean abolishing or reducing evaluation tests but automating them.

ATE is indispensable in evaluating a new family of asics of high complexity. Including the design of a test chip, one engineer needs about half a year for electrical evaluation of one asic family.

4.5 SPECIAL ELECTRICAL PARAMETERS

The following tests are called special, not because they are less important but because they need special skill to be performed. Most of these tests can be performed on some representative members of a family of circuits only. Some kinds of special parameters have to be tested for most families of components. Again, MSI circuits are used for demonstration purpose

4.5.1 Dynamic noise immunity

Section 4.2.2 explained how the static noise immunity can be derived from the measured transfer characteristics. All noise signals below this threshold are innocuous, but not all noise pulses exceeding this threshhold cause false triggering. Short noise pulses are filtered out by the inherent inertia of the affected circuits. The allowable amplitude of noise pulses as a function of their width is plotted in Fig. 4.16(c), using several technologies as examples.

In current mode logic (ECL), the wiring rules for board layouts are restrictive and no noise exceeding the static threshold is allowed. This requires terminated lines of low impedance. No such restrictions are common in TTL or CMOS logic, and terminations are seldom required. Many existing designs in those technologies would not work correctly at all without the relief of dynamic noise immunity. Although this parameter is so important, it is often not explicitely known to logic designers.

Many sudden quality problems in a well-established production line are caused by a decrease of dynamic noise immunity as a consequence of scaling down and dynamically 'improving' the components by a vendor. Dynamic noise immunity is a parameter which is normally not specified by a vendor, because the vendor wants to remain open for scaling down components later. Therefore, it is good practise for a producer of electronic devices to make his own measurements after each design change of the components and to check whether all critical printed lines on boards are safe.

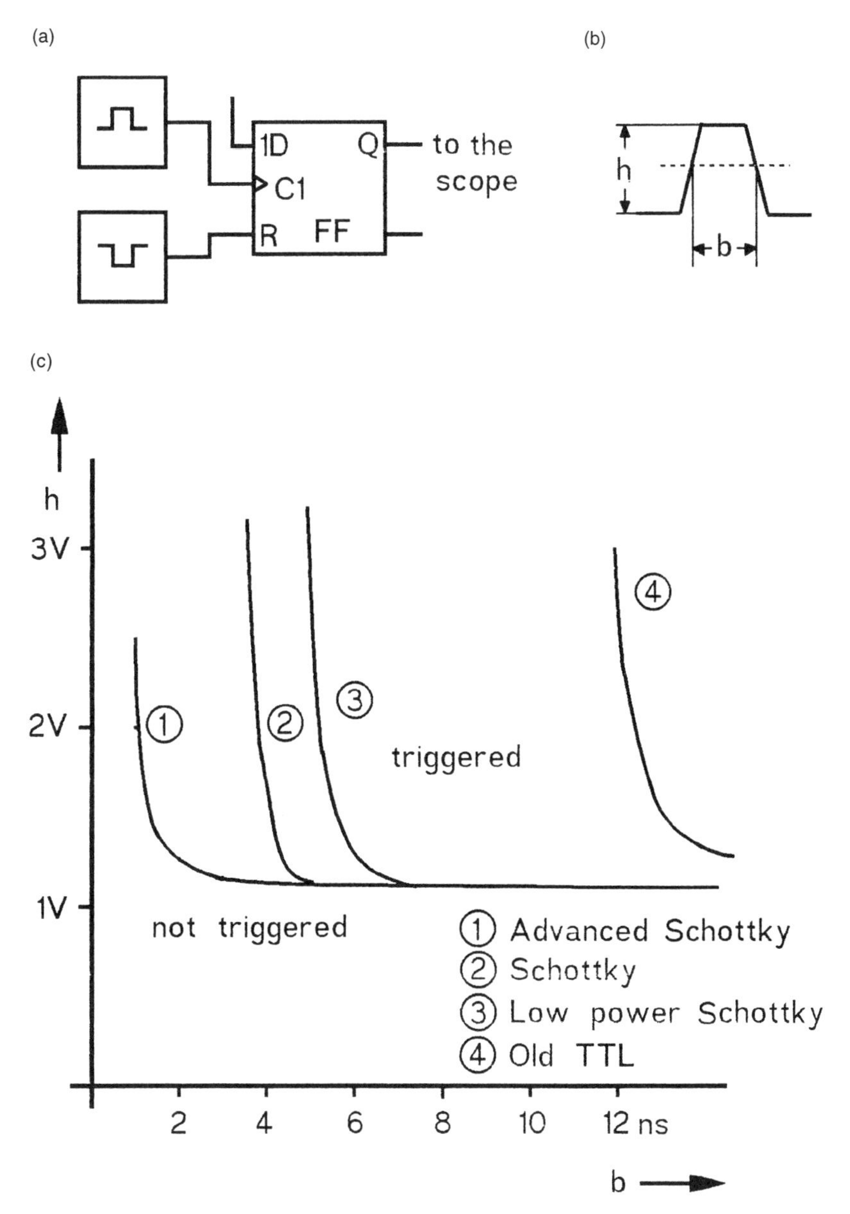

Figure 4.16 Dynamic noise immunity, an important relief to the electrical layout restrictions for logic boards: (a) test circuit; (b) test parameters; and (c) results of measurement.

A method of deriving the allowable safe coupling length from coupling factor and dynamic noise immunity is given in Appendix A. Critical are usually printed lines running parallel more than about six inches with positive going triggering signals. Negative going preset and reset signals and all data signals are less critical.

A high-speed pulse generator with steep edges is used to measure the dynamic noise immunity, either as part of a benchtop set-up or included in an ATE system. A synchronized reset pulse is required for repetitive measurements as shown in Figs 4.16(a) and (b). The DUT can be a combinatorial circuit feeding into the clock input of a flipflop of the same technology or it can be the clock input itself. The result of the measurements, the amplitude of the pulses which are just triggering the DUT is plotted as a function of their width in Fig. 4.16(c). Two items have to be considered:

- Metastability makes the trigger limit somewhat undefined.
- There is no common definition for the voltage level at which the pulse width should be measured. It is best to use the same level as is used for measuring the noise pulses on the line.

4.5.2 Threshold time

A second important parameter to ensure safe board design is the threshold time, i.e. the time for a clock input to stay at threshold level without false triggering. Pulses on long lines on a printed board or on the back plane, will show a step-like distortion caused by reflection (Fig. 4.17(a)). The voltage level of this glitch and its duration depend on the impedance and the length of the driven net. If this glitch is about the threshold level and if its width is sufficient, then the pulse may cause double triggering on the leading edge or false triggering on the trailing edge of an attached flipflop (Fig. 4.17(c)). The threshold time is the admissible width of such a glitch. It depends on the technology and on the design of the flipflop.

In order to measure threshold time, a double pulse is used to simulate such a glitch (Fig. 4.17(b)).

Some typical measured values for the threshold time are:

Advanced Schottky:	5 ns
Schottky:	7 ns
Low-power Schottky:	12 ns

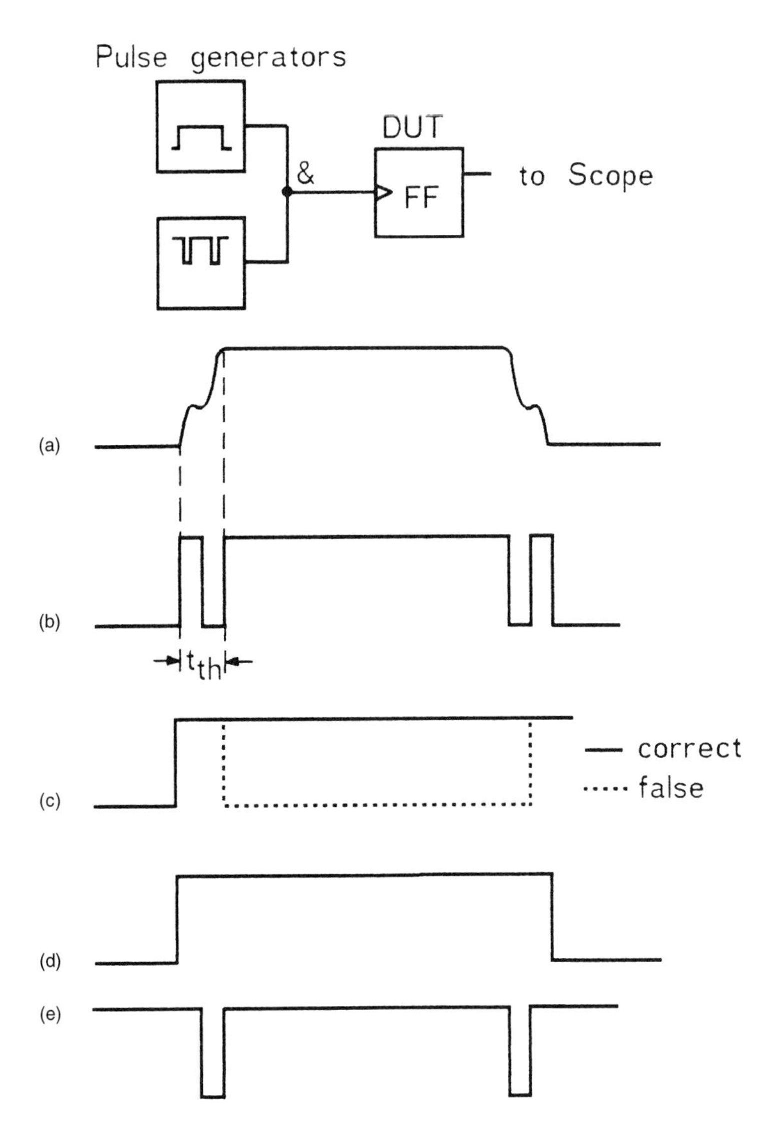

Figure 4.17 Threshold time defines the sensitivity of flipflops to be triggered by glitches: (a) pulse distortion caused by a long line; (b) input pulse applied on flipflop FF to test its sensitivity to pulse distortion; (c) double triggering and trigger at the wrong edge occur when the threshold time is exceeded; (d) pulse shape of the upper generator; and (e) pulse shape of the lower generator.

4.5.3 Dynamic power supply current and supply spikes

Digital electronic circuits with a so-called totem pole output have a peak in their supply characteristics when switching between the logic high and low state. This was shown earlier in section 4.2.4. The reason is that both transistors become conducting, nearly causing a short cut between supply and ground. At the same time, the charge current into the load is drawn from the supply rail. This is the reason for the increase of supply current with frequency for all circuits. It is shown in Fig. 4.18, which is also an example of a vendor's lab data pack. This increase is pronounced for CMOS because of its low static supply current. That means the static supply current is of little meaning for CMOS, the dynamic supply current per Mhz and per pF load has to be specified by the vendor in the datasheet.

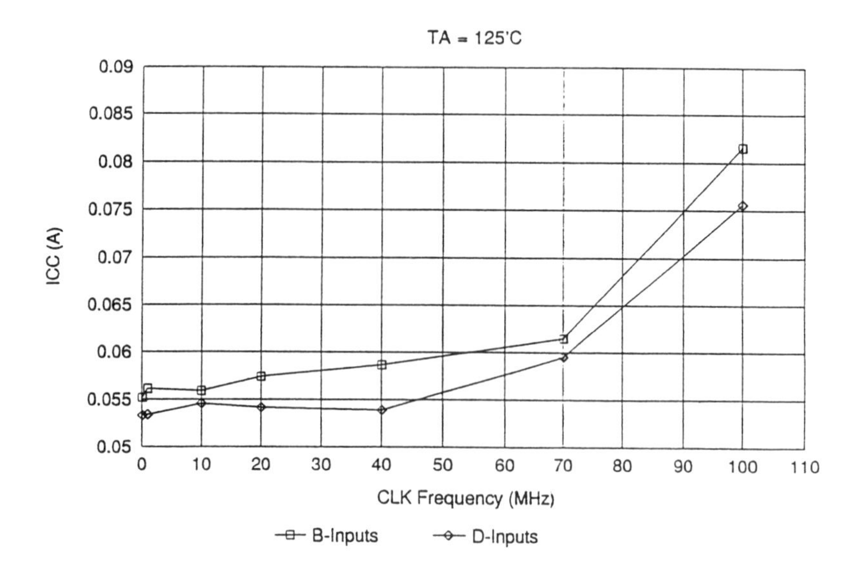

Figure 4.18 Supply current will rise at higher frequency – an example of test results supplied in a vendor's data package.

Not only the increase of current, but also the noise on the supply lines generated by these current spikes, may cause trouble on some applications. This supply current spike can be measured in detail by several methods, e.g. using current probes.

One other method is shown in Fig. 4.19. It is closer to reality, because the generated noise is seen directly. The voltage drop ΔU on a small inductance is measured by a scope. The current can be calculated by

$$\frac{\mathrm{d}i}{\mathrm{d}t} = \frac{\Delta U}{L}$$

With an air coil inductance of 0.2 mmH inductivity, the measured values are 0.8–1.1 V for simple gates and 1.8–2.0 V for power gates.

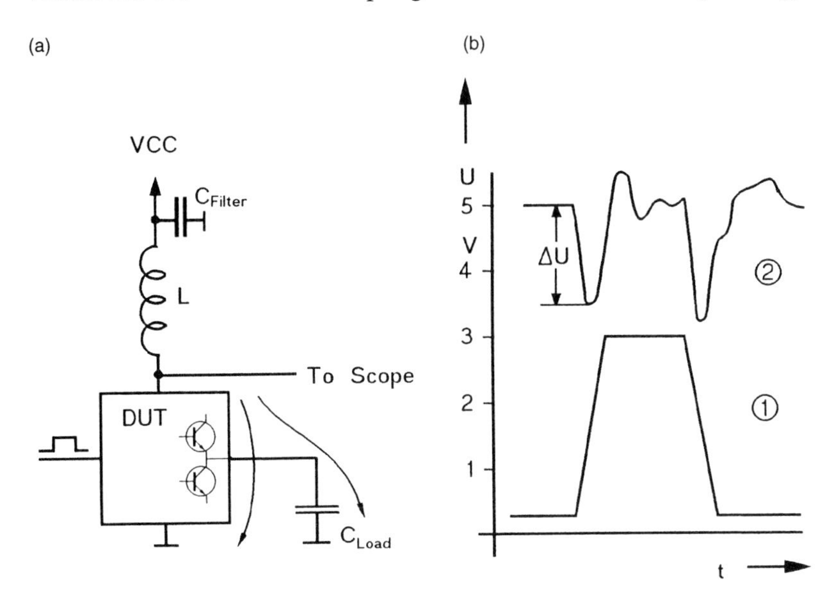

Figure 4.19 Supply curent spike: (a) test circuit; and (b) pulse shape of input voltage (1) and supply voltage (2).

4.5.4 Ground bounce and bus contention

Ground bouncing is a phenomenon occurring inside ICs with several loaded outputs switching at the same time. The sum of the discharge currents is flowing through the internal ground net and the ground pin of the circuit. In theory, this current can be as high as several amperes, at least in the case of fast-switching power drivers. The designers of such drivers are in a dilemma. On one hand, they want to achieve high speed and, on the other hand, they have to ensure easy applicability and that means low ground bounce. Because ground bounce depends on the number of simultaneous switching signals and on the degree of their simultaneity, its effects are often highly sporadic and can cause mysterious failures later in a device. Therefore it is worthwhile measuring this effect on all new components which seem critical, such as all 8–6 bit drivers and latches. The scope drawing Fig. 4.20 is another example for a vendor's lab data pack.

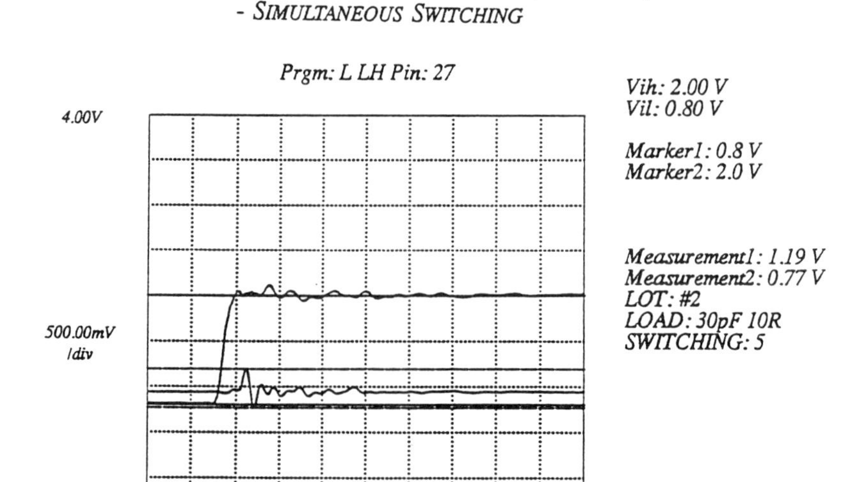

Figure 4.20 Ground bounce, an important design characteristic for bus drivers.

Ground bounce has two aspects:

- A non-switching output in the same package may show a noise spike. If the line connected to this output is long and unterminated, then the amplitude of this pulse will double at the receiving end and may lead to false triggering.
- Flipflops and latches, which are contained in the same package may lose their information by false triggering caused by internal noise.

Similar effects are induced by bus contention. If the outputs of two tristate drivers are connected together to form a wired OR function then only one of these can be enabled at a time. Otherwise, if their output data are of opposite polarity, a high current will flow and produce the same phenomenon as ground bounce. This situation will occur if one driver is enabled before the other one is disabled. To check whether a flipflop or a latch is sensitive to bus contention, the following test sequence has to be executed (Fig. 4.21).

1. Store data.
2. Change input data to the opposite polarity.
3. Execute bus contention (enable EN2 before disabling EN1).
4. Disable EN2 and enable EN1.
5. Check whether the DUT kept its information.

In the case of flipflops, this procedure has to be performed with clock low and clock high during bus contention.

4.5.5 Capacitance of inputs and outputs

Physical basis

The capacitance of inputs and of inactive tristate outputs consists of the package or case capacitance. This is given by the design geometry and by the package material, and by the internal capacitance of the die. The latter depends on the electrical design, i.e. the number of transistors and clamping diodes connected to the pin and their layout. This capacitance is an important parameter because it is responsible for the dynamic loading of the driving output. The dynamic loading may cause an increase in ground noise at the driving gate. Theoretically, a reduction of the driver's lifetime by electromigration might also be possible, but this should not be a problem if the driving output is designed carefully. Dynamic loading also increases the delay time of the driver either by increasing rise or fall time or by a step-like pulse distortion dependent on the relation between edge slope and line length.

Measurement methods

The conventional bridge methods are less suitable for measuring input capacitance because of the resistive and nonlinear parts of this capacitance. The current sense method and the reflection method are

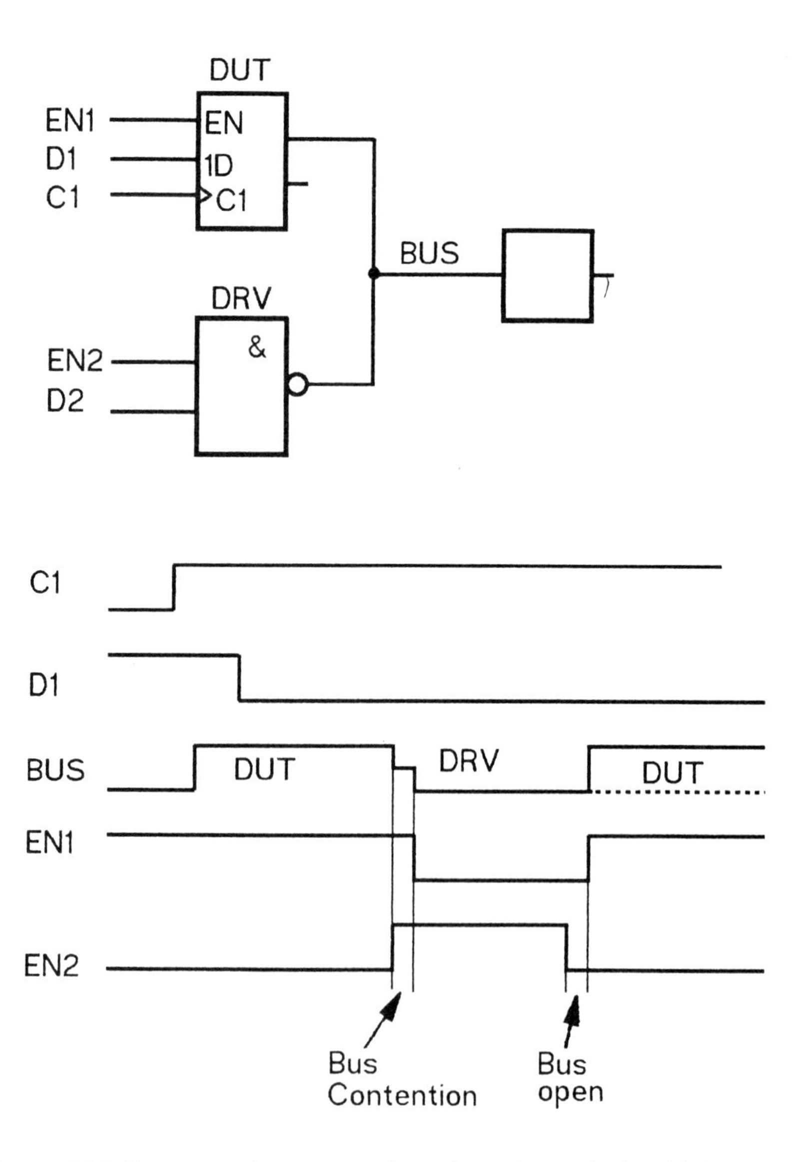

Figure 4.21 Bus contention; to use tristate buses instead of multiplexors will reduce part count and wiring but it requires careful timing to avoid bus contention.

preferable. The current sense method (Fig. 4.22(a)) is based on the fact that the current into the input causes the loading of the driving output. So this current is measured by a current probe connected to an oscilloscope and compared to the current into a discrete capacitor. Such current probes are available as accessories of scopes. A transfer ratio of probe current to scope voltage of 1:5 mA/mV is appropriate. If the switch is situated close to the input then parasitic capacitance will be compensated.

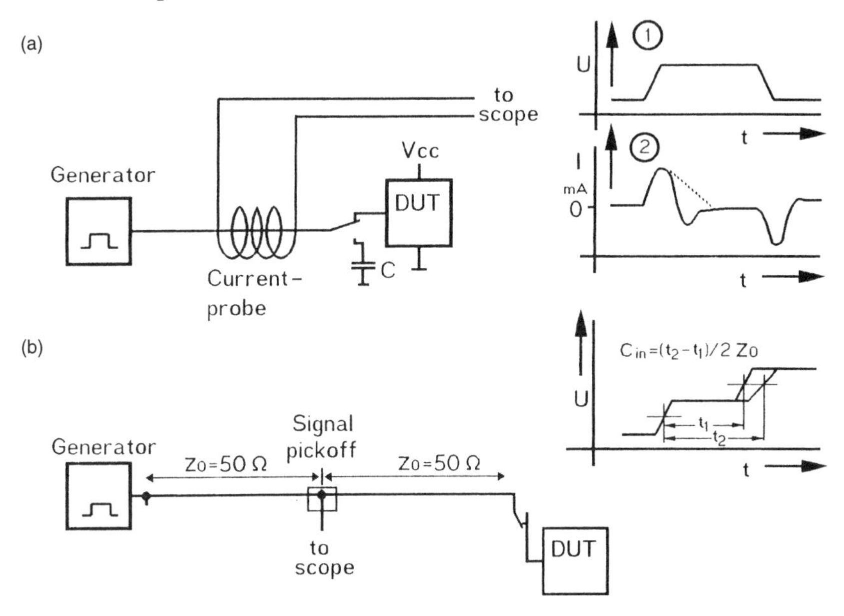

Figure 4.22 Input and output capacitance test circuit: (a) current probe method with input voltage (1) and input current (2); and (b) pulse reflection method with pulse distortion at line input.

Of course, to use a RC circuit instead of a pure capacitor would give a more accurate result. The input and output of an IC is better simulated by a RC circuit. But at least for inputs one cannot find much difference. Static leakage current may produce some error while testing an inactive tristate output (dashed line in Fig. 4.22(a)). The amplitude of the test voltage should be reduced below 2V in such a case.

The reflection method is based on the fact that capacitance will cause a distortion of a pulse on a long line (Fig. 4.22(b)), so this method is useful for high-frequency circuits such as ECL logic.

Results

The total pin capacitance C_{tot} is given by

$$C_{tot} = C_{case} + nC_{in} + mC_{out}$$

where:

C_{case} = package capacitance pin-to-case;
n = number of internal inputs connected to a pin;
m = number of internal tristate outputs connected to a pin;
C_{in} = internal capacitance of an input; and
C_{out} = internal capacitance of a tristate output.

C_{case} can be measured on an empty dummy package to 2–4 pF for DIP packages. C_{in} and C_{out} can be calculated by spice simulation to 2–4 pF and 6–10 pF respectively. Measured values of total pin capacitance for a simple input pin are 4–6 pF and 6–12 pF for a tristate output and up to 18 pF for a bidirectional pin.

4.5.6 Asynchronous behaviour

Physical basis

Each system design engineer will be faced at some time with the problem of synchronizing two digital signals operating at two different frequencies. Even when electronic devices are working internally completely clock synchronous, they are always connected to the outside world by incoming and outgoing signals. Input signals, arriving asynchronously, have to be synchronized to the internal clock. Furthermore, most big electronic devices are combined from several subsystems, like the control unit, memory, peripheral controllers etc., each having its own independent clock. Significant rates of failure result from the necessity to synchronize all signals which link these subsystems mutually. The reason is that, in all synchronizer circuits, the asynchronous signal and the clock are applied directly or through some combinational logic to inputs of a bistable element like a flipflop.

For each flipflop, a minimum clock pulse width and a set-up time requirement are specified. The latter is the minimum time that must exist between data and clock edge so that the data are recognized. In synchronizing two asynchronous pulse trains, flipflops are unavoidably used in such a manner that the specified minimum clock pulse width and set-up time requirements are not always fulfilled. Either marginal pulse widths or marginal set-up times may be formed, marginal pulse widths in the case where the two signals are mixed by some combinational logic.

The word 'marginal' is used to indicate an input condition that is beyond the specified value and which produces an abnormal output state. This abnormal output state may be an oscillation, a slowly rising output, a state midway between low and high, or an increase in the propagation delay. These abnormal output states are commonly called metastable states. The metastable state characteristics of the flipflop used can influence the overall system reliability.

The consequence of metastability is shown in Fig. 4.23. A flipflop FF_1 takes over the data D at time t_1 of clock pulse C. The output of FF_1 is applied to the data input of a second flipflop FF_2 which is triggered at a time $t_1 + \Delta t$. If the input conditions of FF_1 are such that the output comes into the above mentioned metastable state and if the duration of this state is longer than the time Δt, then it is not possible to predict the state of FF_2. Because the duration of the metastable state is distributed statistically, the probability that such a failures occurs can be reduced by increasing the time Δt. But there is no synchronizer circuit that can completely guarantee a faultless response.

This had been deduced from the uncertainty principle of Heisenberg. In a metastable state, the two halves of the flipflop are assumed to be exactly equal, in a half-conducting state and theoretically only one electron decides the final state. Because the impulse and the location of this electron cannot be known at the same time, no certain prediction can be made in advance.

This sychronizer problem has been discussed by several authors. A mathematical analysis of the flipflop settling time and the metastable state mechanism can be found in the literature. Although it is possible to determine the mean time between failure (MTBF) for simple gate flipflops with the help of the given equations, it is quite hard to determine such parameters as amplification factors for more complex ones. For that reason, experimental determination of the rate and duration of the metastable state of commercially available flipflops is found to be more practical (Catt, 1966; Chaney, 1983; Dike and Burton, 1989; Fleischammer and Doertok, 1979; Masseloff, 1993; Kacprzak, 1987; Masteller, 1991; Shear, 1992).

The only relevant parameter design engineers need to know is how long after the specified datasheet maximum delay time they have to wait before using the data, when they have to guarantee a certain failure rate.

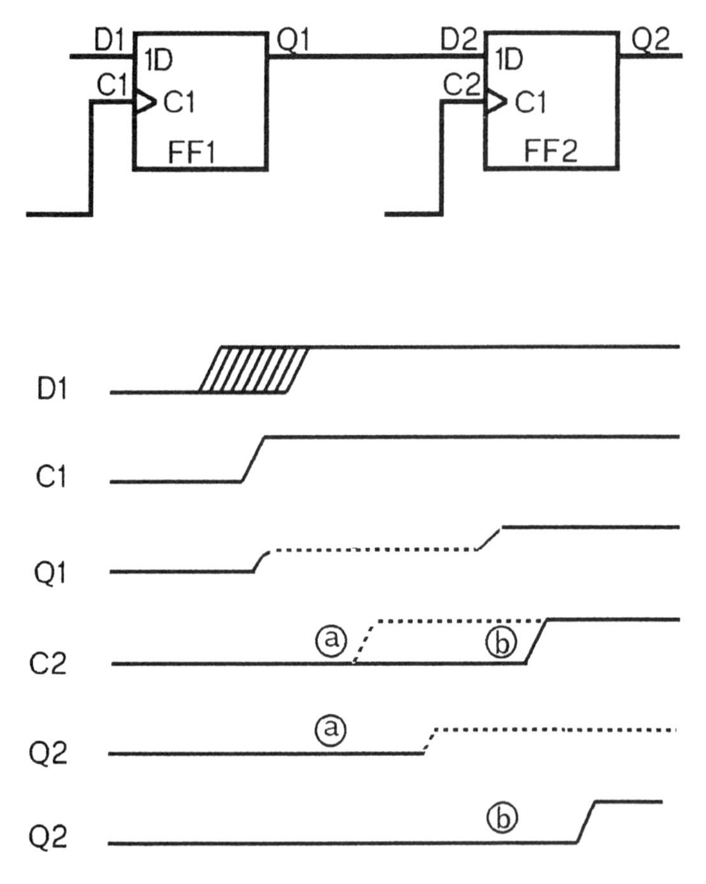

Figure 4.23 The metastability phenomenon is one major reason for sporadic failures, not usually seen on scope. Insufficient set-up time from data D1 to clock C1 will cause metastability at output Q1. The output Q2 will become metastable too (a) unless an adequate waiting time of clock C2 is provided (b).

Test set-up

The reader may conclude, from the above facts, that it is not an easy task to evaluate the metastable characteristics of a flipflop. This is because the number of metastable events is extremely small when compared to the total number of clock transitions and the form of the associated output signals is variable. Only some digital storage oscilloscopes with appropriate trigger facilities allow these signals to be viewed. Therefore they have to be detected and counted by a measuring circuit.

In practice, it is not so difficult to build a special test circuit for characterizing MTBF as a function of the waiting time Δt (time between clock and Q valid). This metastable state-measuring circuit (Fig. 4.24) uses two unsynchronized sawtooth generators (generators 1 and 2), one being very slow and the other fast, which are combined together by a pulse shaper. A differential amplifier circuit like an SN75107 line receiver or an operational amplifier with differential inputs can be used. Pulses of various width are produced at the output (Fig. 4.25) which drive the flipflop under test with marginal pulse widths. The output of this flipflop is connected to two level detectors to recognize metastable edges. Differential amplifiers whose second inputs are referenced to 0.8 V and 2.0 V levels can be used. The strobe facility of the SN75107 enables this comparision to be made at any particular instant referred to a fixed point of time in the system and their outputs are compared by an EXOR gate. Therefore, a metastable state which has the form of a slowly rising output causes a pulse to appear at the output of the EXOR, which is then counted. Other types of differential amplifiers can be used instead of the SN75107 but a separate strobe facility has then to be added. In the case of marginal pulse width, the strobe is derived from the clock edge that is not sweeping.

The circuit described above can be easily converted for the marginal set-up time case by using another generator (generator 3), which is synchronous to generator 2 to provide the data pulse. This is shown by a dashed line in Fig. 4.24. Suitable edges of clock and data are brought together for a specific flipflop and the moving clock edge sweeps the edge of the data pulse and hence produces the marginal set-up time. By interchanging the connections of the two sawtooth generators one can get either the rising or the falling edge sweeping. The strobe has to be derived from the data pulse to be fixed in time and the amplitude of the fast sawtooth should be greater than that of the slow one to avoid simultaneous marginal clock pulses.

For some flipflop types, and for some marginal input conditions, the output rise time remains normal, but the delay time increases. In this case, in which detection of fast rising metastable states is necessary, a small amount of artificial delay is inserted before the differential amplifier which is referenced to 2 V. Otherwise the pulse at the output of the EXOR will be too small to be registered.

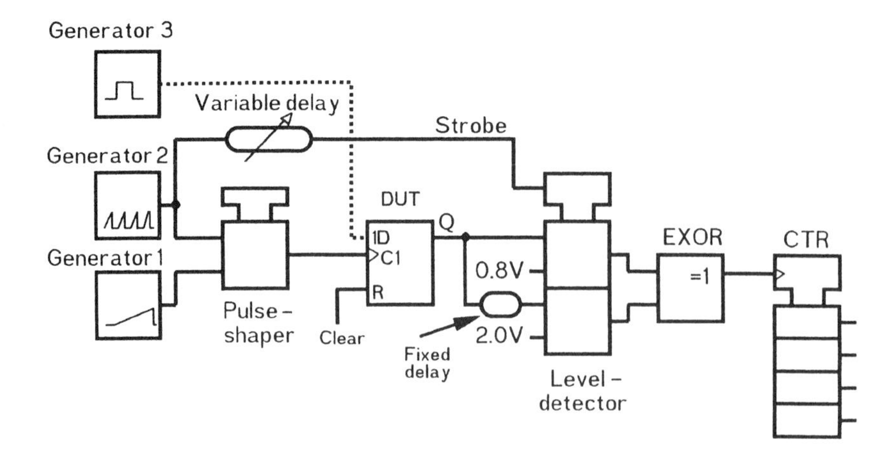

Figure 4.24 Proposal for a metastability test circuit.

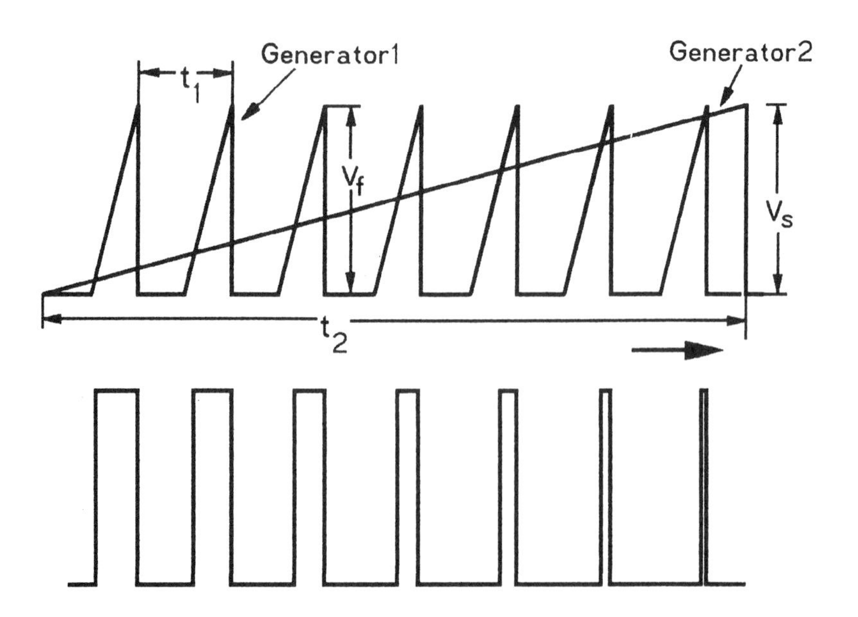

Figure 4.25 How to produce random input pulses; to produce pulses with random width is the key to the test circuit of Fig. 4.26.

Test results

By using the described circuit, the MTBF can be determined from the number of pulses counted in a given time for several values of Δt. This number has to be corrected by the factor $2V_f/V_s$, where V_f is the amplitude of the fast sawtooth and V_s of the slow one, in order to convert to data evenly distributed over the clock period. It can be seen that the failure rate tends to a straight line when plotted on a logarithmic scale (Fig. 4.26). It is obvious that the number of unresolved states depends directly on the clock frequency f_c and data frequency f_D. From the data taken by the above measurements, an equation can be derived which allows the failure rate for different frequencies and different strobe delays Δt to be calculated:

$$\frac{1}{\text{MTBF}} = f_c \times f_D \times \Delta B_1 \quad \text{in case of set-up time}$$

$$\frac{1}{\text{MTBF}} = f_c^2 \times \Delta B_2 \quad \text{in case of clock width}$$

where ΔB_1 and ΔB_2 are constants depending mainly on the flipflop type and the strobe delay Δt. These constants can be determined graphically from the data taken by the above measurements for data and clock frequency of 1 Mhz:

$$\Delta B = k_1 \exp(-k_2 \times \Delta t)$$

the constants k_1 and k_2 depending on the chip temperature, the manufacturer and the manufacturing batch, and on the flipflop type as the main determining factor.

Test results of Fig. 4.26 show that fast flipflop types like advanced Schottky flipflops are less affected by metastability than slower ones. This is valid for simple flipflops and dual rank flipflops only. More complex flipflop designs may show adverse behaviour in spite of high-speed technology.

Conclusions

Metastability was in the past, and is still today, a phenomenon which is underestimated by system designers. Failure rates of several failures per second, minute or even hour are easily detected in a prototype.

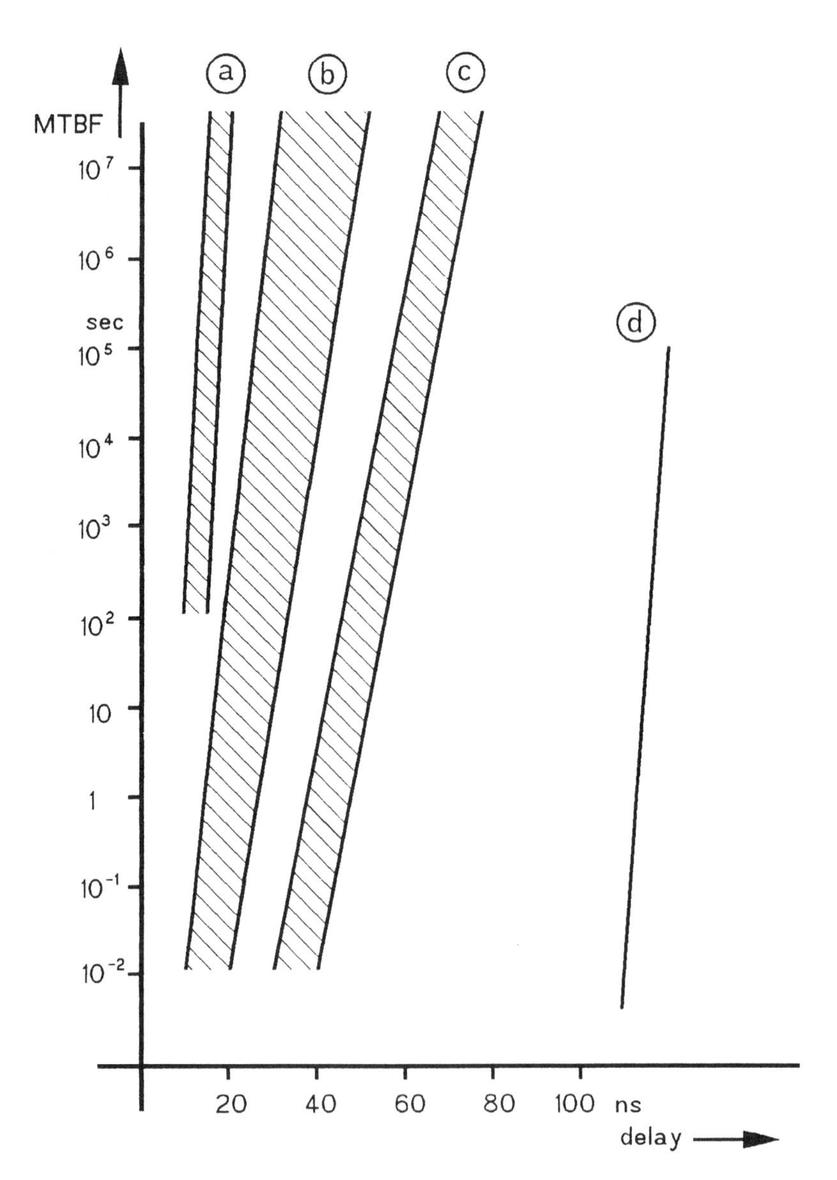

Figure 4.26 MTBF depends on waiting time. The mean time between sporadic failures of sychronizers rises rapidly with longer waiting time. A typical spread of MTBF of simple flipflop types for a data- and a clock-frequency of 1 MHz is shown: (a) FAST and AS flipflops; (b) S and ALS flipflops; and (c) LS flipflops. Note that complex edge triggered flipflops (d) may need a much longer settling time.

But a failure rate of one failure per week – or month – will give much trouble at your customer's site. Furthermore, such failures cannot be detected by simulation. Therefore the quality engineer has to make measuremcnts and give strict application rules to the designer. It is not practical to make incoming tests to choose suitable flipflops. Resistance to metastable failure should be ensured by design. Some vendors offer special flipflop types for which they claim an excellent metastable behaviour, but usually no guarantee is given by them either in the datasheet or in a specification. So the quality engineer should derive his or her recommendations for waiting times from measurements with a safety margin as shown below.

Waiting time t_w = delay time measured by test set-up
(e.g. failures within 10 sec)
\+ max/typ delay time of flipflop (from datasheet)
\+ width of strobe pulse
\+ 20% safety margin for sample variation
\+ 15% safety margin for temperature variation
\+ extrapolation for the tolerated failure rate
(e.g. for one failure within 3 years)

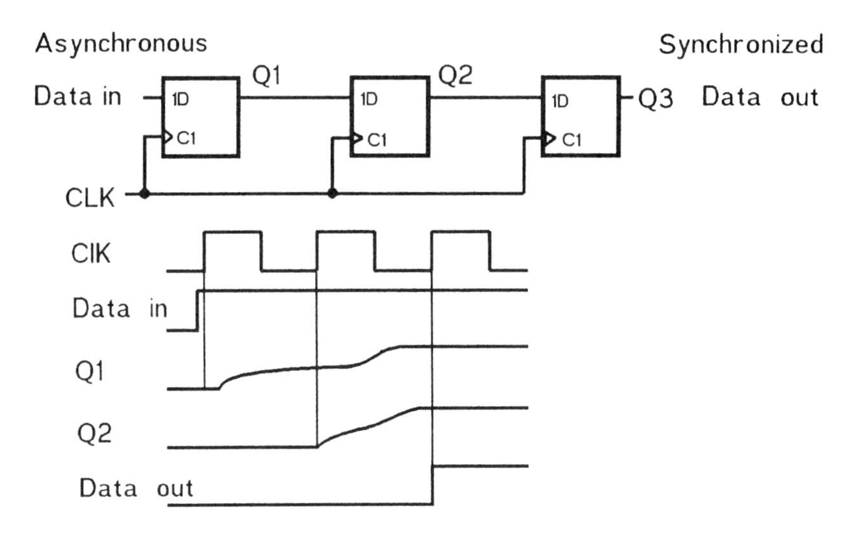

Figure 4.27 A commonly used sychronizer circuit. A two-stage sychronizer circuit makes it easy to introduce delay to realize the required waiting time. It does not avoid metastability and will not reduce waiting time.

An example illustrates this calculation:

$$t_W = 23\,\text{ns} + 5\,\text{ns} + 5\,\text{ns} + 6.8\,\text{ns} + 6.1\,\text{ns} + 8{\times}0.57\,\text{ns} = 50.5\,\text{ns}$$

It has to be noted that the measured values presented in the literature show great differences. This is especially true for the factor k_2 which represents the steepness of the failure rate characteristics.

A practical recommendation for creating the waiting time is the use of a two-staged synchronizer instead of a delay line (Fig. 4.27). It should be emphasized that this insertion of an additional flipflop does not reduce the waiting time; it is only an easy way of integrating a delay of one clock cycle into the design.

4.6 EVALUATION OF STANDARD LSI AND VLSI CIRCUITS (MICROPROCESSORS AND PERIPHERALS)

LSI circuits together with the extensions VLSI, ULSI and GLSI are all called LSI circuits in this chapter for simplicity. These are off-the-shelf circuits containing more than 100 gates in conformance with Table 4.1. They comprise not only all microprocessors, microcomputers, chip sets and peripheral controllers, but also less complex circuits like fifos (first-in-first-out) or some auxiliary circuits. Asics are considered separately in the next section. The main difference between LSI circuits and SSI/MSIs is not the number of gates, though this is generally assumed as listed in Table 4.1. The real distinction is their quite different logical structures. SSI/MSIs are logically defined by some kind of truth table. The state of the outputs depends on the inputs and on the outputs of the preceding state (state machine), as explained in Fig. 4.28.

Although the formula of Fig. 4.28

$$\text{Out}\,[n] = f(\text{In}\,[n], \text{Out}\,[n{-}1])$$

is valid generally, it is not very practical for LSI circuits. Their function is far better defined by a literal description, as is given in the vendor's datasheets. The physical reason is that LSIs usually contain some kind of microcode stored in an internal ROM. Once the entry point is given by an input of a command word, this microcode defines autonomously the sequence of states.

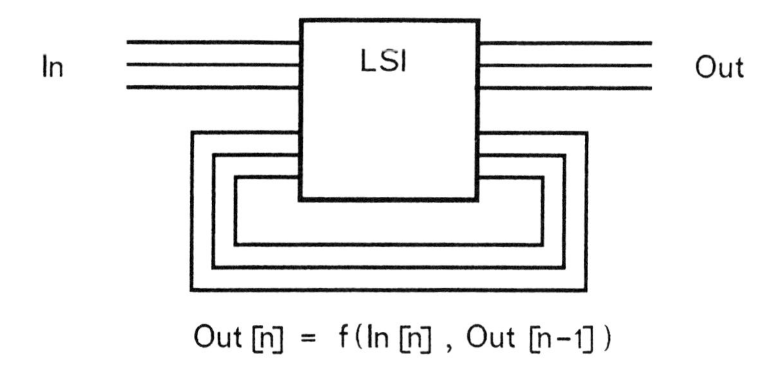

(b)

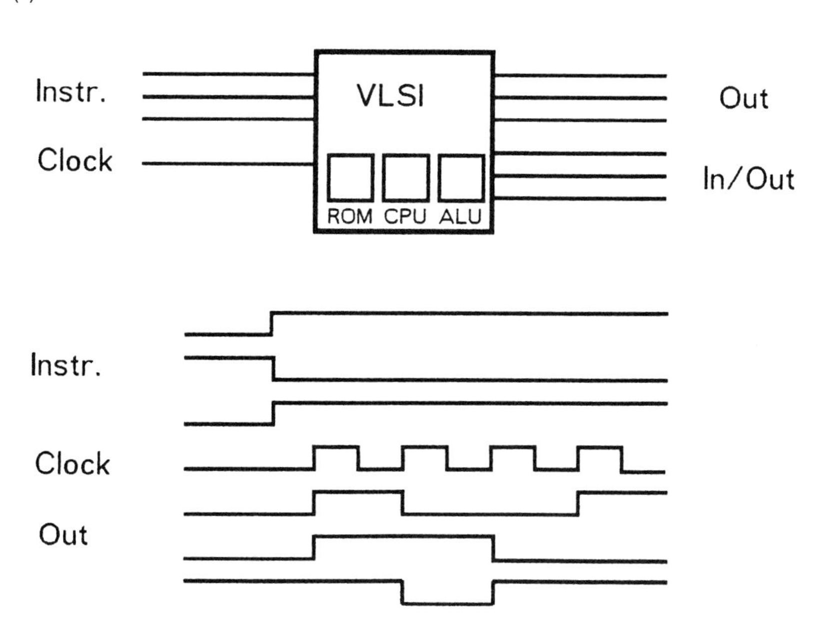

Figure 4.28 VSLI circuits, like microprocessors, began a new era in the world of ICs: (a) in practice LSI circuits may be treated as somewhat larger MSIs; (b) VSLI circuits are different.

This fundamental difference in functional description gives a different way of evaluating these components. Static and dynamic parameters are similar to SSI/MSI, as are the associated test methods. Although static parameters refer only to the few gates at the border of LSI, testing these also helps to characterize the internal technology. The difference lies in the functional test, which is much more significant for LSIs than for MSIs. Again, it is the quality engineer's first task to understand completely the functional description given in the vendor's datasheet. Of course, the vendor's application engineers try to inform their customers as accurately as possible by means of documentation and training courses. However, the complexity of the functions will still leave many uncertainties which should be cleared away by discussions with the vendor. This is especially true for complex instruction set computers (CISC) and for peripheral controllers. But for a really deep understanding of the functions, the customer's quality engineers should do some tests of their own (Lombardi, 1988; Reghbati, 1991).

Microprocessors are submitted to intense simulation by the manufacturer and are tested thoroughly before they are offered for sale. Nevertheless new microprocessors are by no means bug-free. Some bugs are detected a long time after the first delivery of new software. The advantage for the user lies in the large customer base affected by such bugs. The vendor will be highly interested in clearing up all problems quickly. More difficulties are to be expected when using the appropriate peripherals. Often they are designed soon after the processor has been introduced and are released in a hurry so as not to lose time-to-market. Final debugging is left to the customer. The customer's quality engineers have to analyse all problems arising, whether they are application problems or component bugs, and discuss them with the vendor.

The reasons behind tests carried out by the customer are similar to those explained in section 3.1. The quality engineer has to consult the management when selecting a LSI circuit which will best fit the needs of the specific application. This is a decision of great significance for the company. It is one of the management team tasks, as mentioned in sections 2.2 and 2.3. The quality engineer also has to:

- give advice to the logic design engineers about application problems with the component;
- be a link to the manufacturer's design group to obtain competent information beyond the field application engineeers;
- be able to analyse unexpected application problems occurring

later in the prototype phase;
- find initial solutions to overcome such problems until a design change becomes effective; and
- be able to select immcdiatcly good samples as a first remedy for quality problems in the production phase.

The quality engineer also has to assist the design group in logic simulation of devices containing LSI components. Usually no software simulation model is available for VLSI circuits, so a hardware model has to be provided in time and integrated into the simulation process. A simulation of the complete electronic device including all asics, VLSI, MSI and SSI components is desirable. It will not only reduce time-to-market considerably but also improve the quality of the product as mentioned in section 2.2. This is because the device will enter the market with less hidden design errors. He has to:

- procure test programs and loadboards for incoming testing as long as no ship-to-stock procedure is installed; and
- locate a second source and check the functional differences between it and the first source.

Because of the high complexity, manual tests are not recommended. A benchtop test system consisting of a powerful word generator and a high-speed logic analyser has to be used. This analyser should be able to detect possible glitches and noise pulses. The advantage of this test system is the ability to simulate the real usage of the component at real speeds. Adverse effects of application programs can be detected while parts of these programs are running on the system. The disadvantage is that the transfer of input and output pattern from the logic analyser to a test system for incoming inspection may be difficult because the analyser is event-triggered whereas the tester is clock-triggered.

To overcome this difficulty, the second method uses an automatic test system as an analyser. The method is then as follows.

- Use the timing data on the vendor's datasheet to program the tester's timing system.
- Input the commands you want to test from the command table on the vendor's datasheet with dummy data at the output pins.
- Put the tester in a learning mode. In this mode, the tester compares the output of the DUT with its own dummy output data. Then it changes the dummy data according to what it learned from the DUT and repeats the test step. So finally it has learned the response sequence of the DUT.

- Analyse the obtained result and compare it to the functional description given by the vendor.
- Repeat this procedure for all commands under critical conditions. Assembler code can be used to simplify the input of larger command sequences.

By this procedure each command can be evaluated stepwise in detail and all possible doubts cleared away. Optionally, all these procedures can be integrated into one evaluation test program which can be used to evaluate easily all redesigns and second sources. You can use this evaluation test program for incoming inspection too if no test program is available.

Using this method, it is not difficult to list the output states during internal operation of complex commands (e.g. multiply) and to find the number of clock cycles necessary.

Some LSI circuits have so-called undefined or secret commands, i.e. not all command words are defined in the datasheet. A well-known historical example is the 8-bit processor 8080. Only 250 of the possible 256 commands had been specified by Intel. The behaviour of the six unspecified commands was evaluated by users.

Such secret commands are used by some manufacturers to put their circuits into a test mode, which allows them to perform outgoing inspection more easily. Many peripheral circuits, e.g. disc or tape controllers, have subsections with widely different timing, one section with high speed to interface the host and another with low speed in the range of several milliseconds to control the mechanical parts. Those components would be very difficult to test without special precautions such as a test mode. This is not normally mentioned in the datasheet, so ask your vendor and check it yourself.

A priority in evaluating LSI-circuits is determining the range of correct functional operation. The results of this tests are displayed mostly by two-dimensional plots with pass–fail as a third dimension, for instance:

Supply voltage – ambient temperature
Input low level – input high level
Operating frequency – supply voltage

Such Shmoo plots give interesting information, which is also relevant to the quality of the end product, if holes, gaps or notches show up in the graphs (Fig. 4.29). Unexplainable failures will occur if the device runs into such a condition during operation.

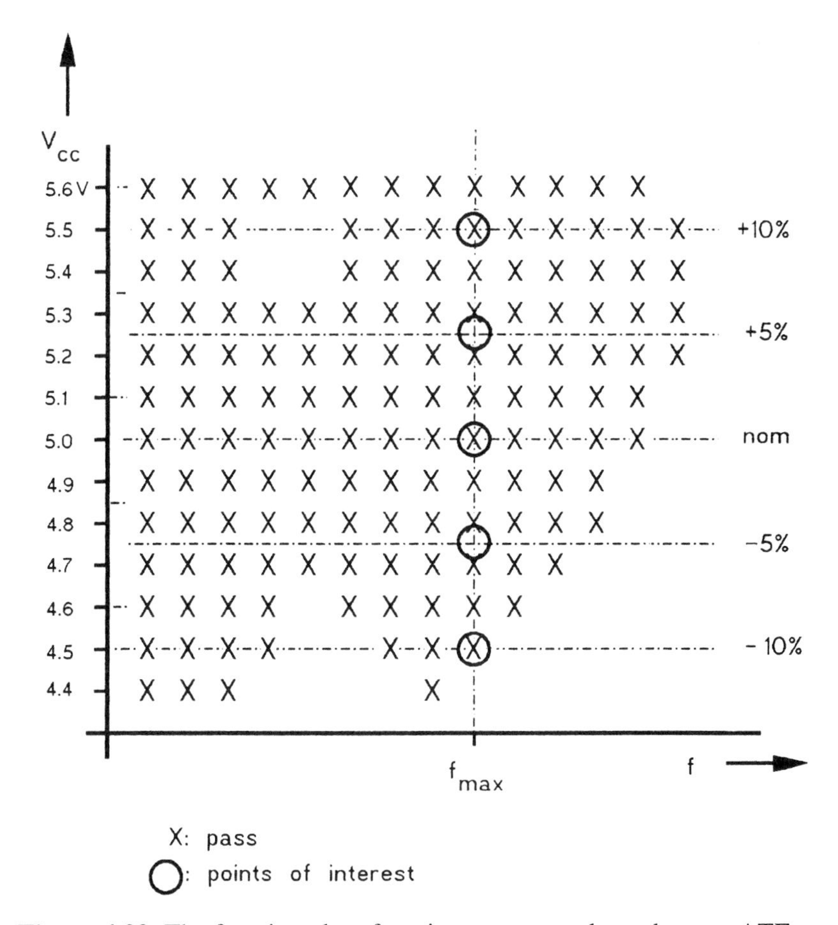

Figure 4.29 The function plot of a microprocessor shows how an ATE tester can support evaluation. Holes in the function plot of VSLI circuits may indicate a potential reason for sporadic failures.

Metastability is also a vital item which has to be considered by the designer when using LSI circuits. Ready and interrupt inputs to microprocessors have no fixed timing relation to the processor clock and also the request inputs to a multimaster bus arbitration. Fortunately you do not have to expect significant problems with popular microprocessors. Most silicon designers take potential metastability problems into account and build in two-stage synchronizers. The same is true for multibus. There is no general rule, however. Some dual-port memories and asynchronous fifos need external synchronization, which is not mentioned explicitly in the datasheet. Full and empty conditions are exposed to metastability; therefore, some fifos have nearly empty and nearly full signals which allow the designer to avoid this condition. It seems to be good advice for quality engineers to perform metastability tests as described in section 4.5.6 during the evaluation of LSI circuits, if these are used to synchronize data to an internal clock. If metastability problems show up later at a customer, the results will be costly and time-consuming.

The lack of second sources is a problem for many VLSI and ULSI circuits. This is less true for commonly used microprocessors. It is more a problem for special ULSI circuits with a relative small customer base. The best advice is to build up a store of these parts to protect against sudden discontinuation of delivery and to prepare by designing a work-around.

4.7 ELECTRICAL EVALUATION OF ASICS

Asics are of growing importance for the design of all kinds of electronic devices. In Fig. 4.1, one example was given of the effects on the design of electronic devices caused by the use of asics. System integration, the computer together with peripherals and first-level memory on one chip, is one goal. The impact of asics is one of the reasons for the digitalization of traditional analogue designs e.g. telecommunications and audio/video systems. Therefore, this subject is treated more explicitly in this book.

4.7.1 Different categories of asics

An overview of the different categories of asics is given in Fig. 4.30. The problems which arise from evaluation and use of asics are quite different for each kind. They are described in the following sections.

4.7.2 Full custom asics

Full custom asics are designed at transistor level in close co-operation between user and vendor. They can be seen as special LSI circuits designed and produced by the vendor at the request of the user. Evaluation can only be performed as part of a special agreement and is not considered in this book.

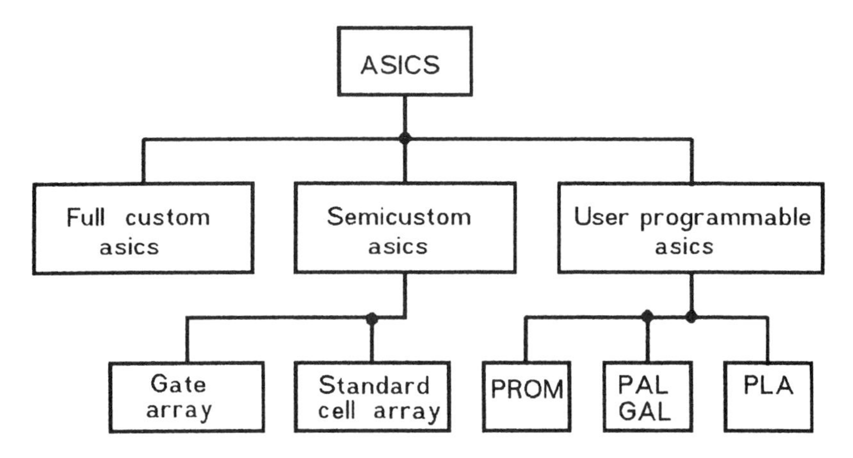

Figure 4.30 How asics can be subdivided into categories.

4.7.3 Semicustom asics: gate arrays and standard cells

Design quality into gate arrays

Gate arrays are made from prefabricated wafers, so-called masters, which contain rows of gates or a sea of gates, by adding individual metallization layers. In standard cell arrays, the masters are individually composed from prerouted standardized cells, which are selected by the user from a cell library. But today the difference between both kinds of asics vanishes; modern asics, sometimes called 'customer-defined arrays' or 'flex-arrays', contain regions for standard cells, e.g. RAMs, ROMs and processor cores, and regions for gate arrays which contain random logic. So they are both called gate arrays in this book.

Asics are not merely a replacement of traditionally designed standard logic in order to reduce board space, to increase reliability, to improve performance, and so hopefully save cost. Using asics also means a change in the design flow on the user's side and a change in the design management. The user can expect the same problems which arise when using standard logic as well as many additional problems which emerge from the close interrelation between vendor and customer, imperative to the achievement of a good design. A lot of information has to be exchanged between both parties, as shown in Fig. 4.31. In this figure, only an overview of the design procedure of modern asics is shown. A detailed presentation is beyond the scope of

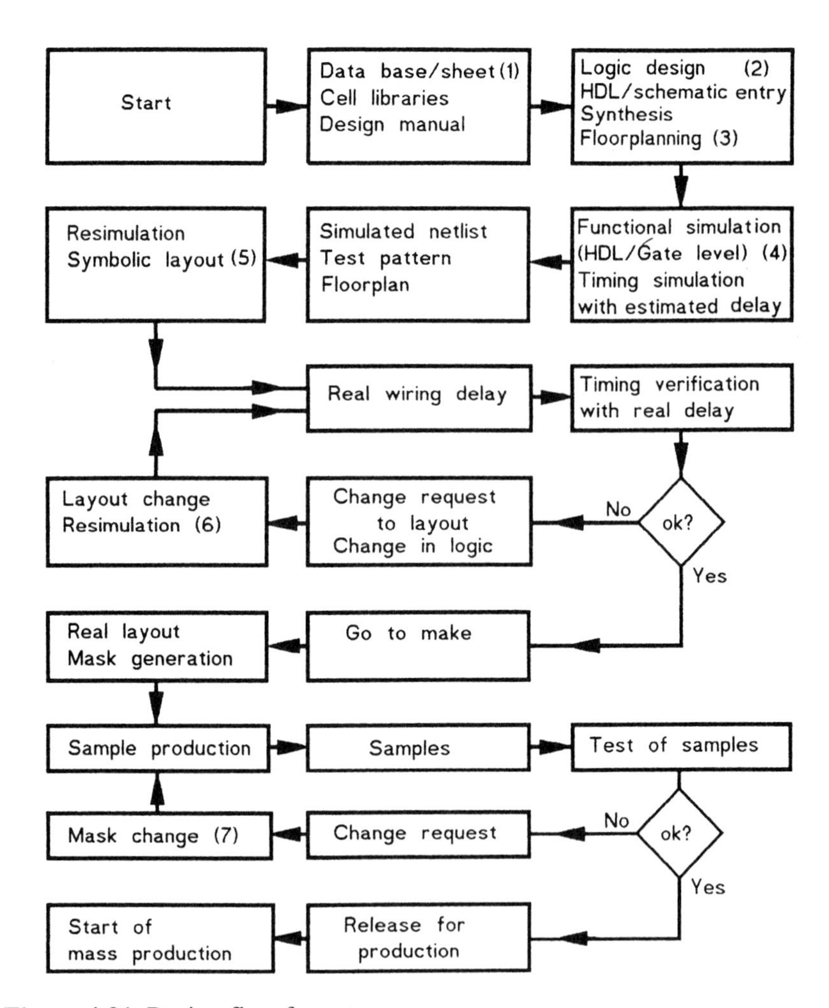

Figure 4.31 Design flow for gate arrays.

this book. The notes in Fig. 4.31 hint at recent improvements of the design process.

1. A software database is used instead of a printed datasheet as an information medium for the customer. This is easier and faster to update, and necessary because with advanced asics the final tuning of technology by the vendor and the first designs by the customer may overlap.
2. The former graphical schematic entry on cell level is going to be

partly replaced by HDL (high-level description language) entry with subsequent logic synthesis. Verilog or VHDL are examples for such tools.

3. It is recommended that customcrs do some floor planning. This will allow him to reduce the delay values of critical paths and reach a more exact estimation of delays.
4. Functional simulation without timing constraints is done first. As well as simulation at gate level, simulation at HDL level is gaining more attention because the simulation time is reduced considerably. The timing simulation with estimated delays is the second step where the principle of 'static timing analysis' is often used.
5. Symbolic layout is an intermediate step without regard to line width and distance.
6. Two corrective cycles are shown: (6) before and (7) after go-to-make.

During such a transfer of data, many open questions arise on both sides and much misunderstanding. Each pass of data will not be a single shift from one side to the other but most often a sequence of cycles. Questions and answers, explanations and meetings, errors and retries will occur before a co-operation is well established. In the end, the customer's design engineers may become frustrated and unwilling to do further designs with this vendor. But changing the vendor will not solve their problems; they only will have to run through the same learning curve once again. It is the duty of the quality group to smooth the way for the logic designers. A joint evaluation with the vendor will be more beneficial than in the case of standard circuits.

First, you must choose the right asics vendor. This is far more important, as any later change will mean a redesign and a considerable delay for the user. As will be shown below, it is possible to change a vendor at a later stage of design, but this always entails some risk and requires a design which makes provisions for such a possibility. So one very important point is the long-term trustworthiness of the vendor. Such a major strategic decision can be made only by the upper mangement. An agreement must be reached between the logic designers and the quality manager (Chapter 2). The quality group's input can only come from a study of the vendor's data. A questionnaire to guide through this study is given in Table 4.3. This questionnaire points to the main requirements on asics which are generally valid but each customer may have its own special requirements. In Table 4.3 examples are listed designated with

Table 4.3 Questionnaire to compare vendors of asics

Item:	*(Note)*	*Vendor A*	*Vendor B*	*Vendor C*	*Vendor D*
Technology:					
Channel length (μm)	(1)	1.0	0.8	0.5	0.35
Number of metal layers		2	3	3..4	5
Gate count:					
Maximum number		160k	400k		
Usable gates	(2)	80k	300k	800k	1200k
Chip dimensions:					
Length × width (mm)		15×15	15×15	19×19	19×19
Package:					
Pin count total		320	540	652	836
Pin count logic		280	410	480	518
Delay time (ns)					
Gate delay typ.	(3)		0.4	0.3	0.2
Factor max/typ	(4)		1.8		
Mix delay max.	(5)				
Supply current:					
I_{DD} (μA/Mhz/gate)		6.5	3	1.3	0.6
Output drive current:					
Normal output		3	4		
Power driver		6	8		
Simultaneously switching/V_{SS} pin	(6)	12	16		
Cell library:	(7)				
Number of cells			150	falling	
Number of megacells			40	rising	
Number of soft macros			100		
Megacells:					
RAM size max.		40kb	144kb	900kb	1.8Mb
RAM access time (ns)		15	10		
ROM size max.		64kb	1Mb		
Data path generator	(8)	yes	yes	?	
Processor core		yes	yes	yes	yes
Price:					
Price/unit					
Development cost					
Turnaround time:					
Before go-to-make	(9)				
After go-to-make	(10)				

Notes for Table 4.3:

(1) Drawn gate length.
(2) Usable gates in case of a design with gates only (no RAM or ROM).
(3) Maximum of $\{t_{phl}/t_{plh}\}$ for $F_O = 2$ and typical conditions.
(4) Tolerance factor max/typ. Maximum value, worst case condition.
(5) Maximum delay for a customer-defined mix of cells.
(6) Number of simultaneously switching normal outputs allowed per ground pin.
(7) Cells or hard macros with wiring fixed by design megacells are RAM/ROM/processor core, etc. soft macros with wiring defined by router.
(8) Not much emphasized.
(9) See point 6 of Fig. 4.30.
(10) See point 7 of Fig. 4.30.

vendors A–D. These examples are given for the conveniance of the reader only, to show a rough estimate of what figures he or she can request. Vendor A means older technology from about 1990, vendor B, the technology of 1992, vendor C, 1994 and vendor D, what can be expected in 1996. For 1998, a technology with 0.25 μm gate length and 23×23 mm chip dimensions is said to be planned.

The second step is an evaluation of the gate array. The problem in evaluating gate arrays lies in the fact that usually no hardware is available at the evaluation phase. The first silicon comes only at production phase, which may be too late. A test chip is necessary to evaluate gate arrays in time. The evaluation results must be available before the first silicon is ordered. Seen from the point of view of cost, the test chip is at best supplied by the vendor in the course of a joint qualification. Often the vendor will supply only a so-called technology chip containing a few gates to test the speed of the technology in a small package. This chip is absolutely necessary for the vendor and for the customer to prove that the technological goals has been reached, but it is insufficient for a complete evaluation as listed below. In this case, or, if no test chip is available at all, then you as the user have to design the test chip. You can get some price reduction by sharing the measurement results with the vendor.

Whether an evaluation of gate arrays is appropriate depends on:

- the number of designs the user intends to make with this vendor and technology;
- the vendor's experience – whether the technology, which the user has in mind is new, or is established for the vendor; and

- whether the designer uses the technology at its utmost speed limits or leaves a safety margin.

To use a more established technology would spare the test chip but eventually result in a less competitive electronic device. Generally, most designers think that to use and to evaluate a more advanced technology is the better way. Relatively few designers are shy of the risks (and eventual costs) and prefer to use less progressive designs with a more established technology.

Given as an approximate value, Fig. 4.32 shows the number of redesigns necessary for logic errors and also for technologically caused errors as a function of the complexity. From several design examples, we can conclude that for complex gate arrays, i.e. more than 50 000 gates, 80% require at least one redesign, 20% more than one. So, for each design, one redesign has to be calculated; input/output devices may need less, processors with strong partitioning more For smaller gate arrays, with less than 10 000 gates, 20–30% of the designs have to be redesigned. These figures are valid in cases where a test chip has been made beforehand. At least one additional redesign is necessary without a test chip, if an advanced technology from a new vendor is used. Therefore, it is strongly recommended that a test chip is evaluated, if more than 3–5 designs are to be made in a new technology by a new vendor. This is not only to save cost on redesigns, but above all to save time (Comerford, 1989; Eichelberger and Williams, 1973; Marshall, 1993; Totton, 1985).

Test chip

The first thing to do in designing a test chip is to make a detailed plan of what parameters are to be tested and how. From this an overall structure of the chip – a floor plan – has to be derived. Some users use a preliminary version of their planned logic design as a test vehicle,and when their design is completed, the results of the test chip are available. However, this has limited benefit as only a few of the questions described below can be answered.

The test chip should preferably contain special logic designed only for test purposes. One very important restriction for the logic of such a test chip is the fact that it must be testable in the same way as a normal logic chip by an automatic tester. So, any special structures like direct connection from the pins to the internal cells must be avoided. A test chip is more pin-limited than a normal logic chip. Providing as many test structures as possible with the given number of

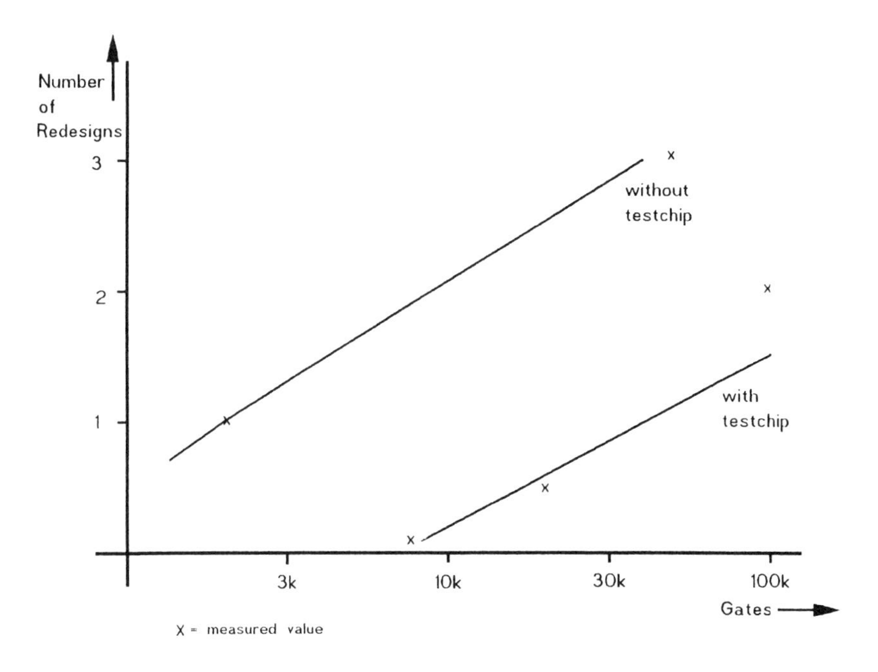

Gate count	*Designs*	*Redesigns*	*Test chip*
2k	5	5	No
8k	17	2	Yes
20k	7	4	Yes
50k	8	22	No
100k	5	10	Yes

Note that all asics used advanced technology at the time of design.
A test chip will reduce the number of redesigns of asics and so reduce time to market

Figure 4.32 The number of redesigns of asics.

inputs and outputs can be achieved by the principle of multiplexing. This principle, using 8:1 multiplexers, is illustrated by a block schematic in Fig. 4.33. An output mutiplexer 1 connects the individual test structures 8 through the output driver 2 to the output pin 3. On the input side, all test structures are driven in parallel from input pin 5 through input driver 6. A demultiplexer 4 can be used optionally to switch off unused test structures, which is recommended. Both multiplexer and demultiplexer may share their address select lines 9. So, the number of input and output pins is reduced considerably. This

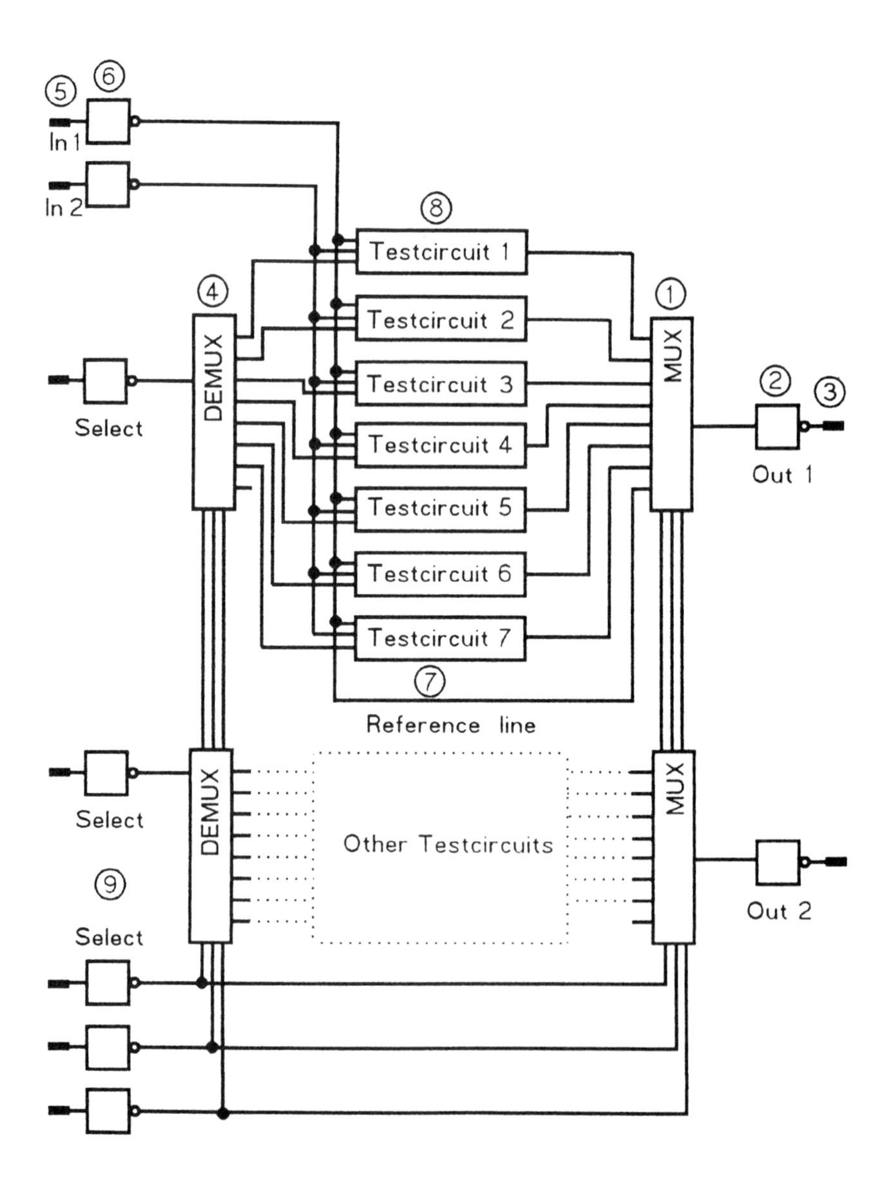

Figure 4.33 The design of a test chip. The logic depth of a test chip is low, the chip is pin-limited. Multiplexing inputs and outputs is the way to obtain maximum information from a test chip.

method is best suited for rather small test structures like SSI and MSI circuits. The test circuits are selected one after another by proper signals on the select pins. The following proposals detail which parameters have to be measured by the test chip.

Logic function of some important cells

If a vendor introduces a new family of gate arrays in a new technology then usually not all cells are tested in hardware by the vendor, especially if it provides a large library containing new cell types. A great many of these cells are only simulated. These cells may contain technological failures, like forgotten through-holes. A mere shrink of an existing technology will bring less problems, except with input/output drivers which have to be redesigned when shrunk. In spite of the fact that more and more cell designs are generated automatically, this verification is advisable for all cells used by a customer for the first time.

Dynamic performance

The dynamic parameters from the vendor's datasheet may not be sufficient to do an excellent design; some parameters may be lacking, others come from simulation and not from measurement. Examples of lacking parameters are: clock pulse pause ratio, distance from clear to clock and loaded delay. In general, the same parameters listed in section 4.3 for MSIs also apply to asics. Special attention has to be paid to the test conditions in the vendor's datasheet if the customer uses the technology at its speed limits. In order to eliminate any doubts, careful delay measurements by a test chip are recommended. This may become a major cost-saving benefit of the test chip.

Because a direct connection from pins to internal cells is not recommended, two test structures are usual to make delay measurements on cells which are inaccessible inside the chip: either ring oscillators or a chain of gates. In Fig. 4.34 logic diagrams of such test structures are shown.

Using ring oscillators is the most exact method to define the pair delay of gates and thus the speed of a technology. To improve the accuracy of measurements, several cells of the same type are connected in series. One oscillator ring composed of only one type of gate should be taken as a reference. This is presented in Fig. 4.34(a) for two input NAND gates.

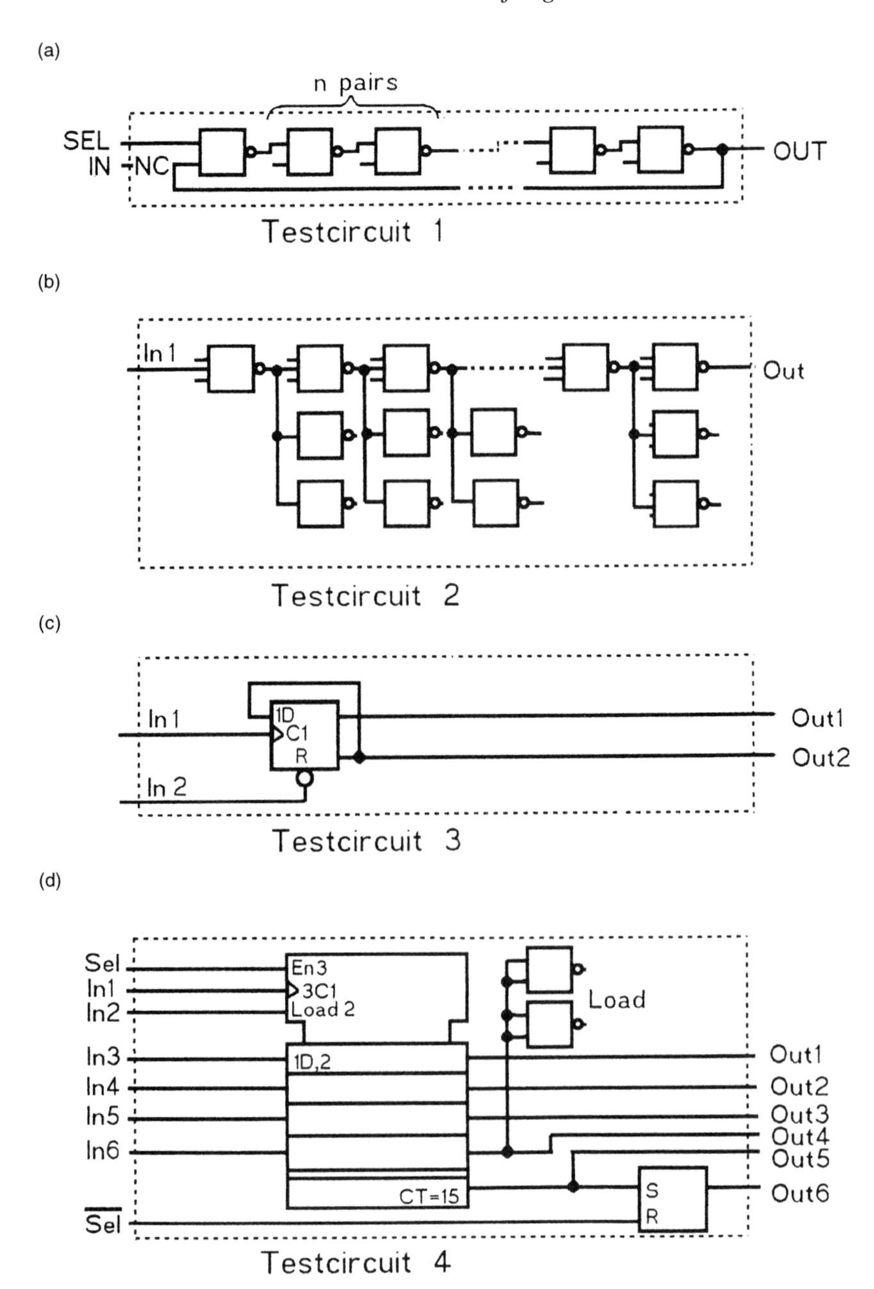

Figure 4.34 Four examples of test cells to check the dynamic behaviour of a gate array by a test chip.

A chain of gates is used to measure rising and falling gate delay separately for different kinds of gates (NAND, NOR, etc.), different loadings and different fans. A comparative measurement method has been established, as shown by Fig. 4.33. A direct connection, 7, between demultiplexer and multiplexer is used to measure the delay of input and output cells alone. This delay value is then subtracted from the measured value from input to output pin of the other test structures.

Figure 4.34(b) gives the logic diagram of three input NAND gates loaded by two inputs each. Flipflops, adders, counters and other MSI cells can be tested in the same way. One important parameter of a flipflop is the maximum toggle rate, as shown on Fig. 4.34(c). All flipflops must have provisions to be put into a definite status at initialization. This is also because the test chip must be testable by an automatic test system. Static preset or clear inputs will fullfil this test restriction. Figure 4.34(d) shows that more input and output multiplexers are necessary to test complex MSI cells like counters. In the example shown, one output is loaded. This asymmetry can lead to sporadic noise pulses at the carry output. A latch flipflop at the carry output will store such possible noise pulses.

Test of standard cells

Special test circuits have to be made to test standard cells containing RAM, ROM or microprocessors. The boundary scan technique is sometimes proposed for this purpose. The RAM block can be isolated logically from the rest of the logic and tested separately with only a few additional pins. Another alternative is the use of built-in self test (BIST). A BIST compiler should be available from the vendor. A good compromise is to connect all address inputs and clocks to external pins via multiplexers and to use a boundary scan for data only.

Asynchronous behaviour of flipflops

Here the same parameters have to be tested as in section 4.5.6 and the same test set-up can be used. It is not necessary to implement the detecting circuit on chip because the metastable signals propagate through the output drivers. The circuit in Fig. 4.34(c) is suited to this test.

Ground noise

Ground noise occurs when many outputs are switching simultaneously. Figure 4.35 shows an example of how to implement test logic. This test determines the necessary number and assignment of ground pins. A larger number of ground pins may be provided in the test chip and partially connected in several configurations to the ground of the test board to find the necessary number. Input pins leading to clock inputs (4) adjacent to outputs from wide data buses are critical. Such critical inputs have to be shielded by adjacent ground pins (8). For large asics, two or three wiring layers on-chip are state of the art. A ground net or at least a ground ring connecting all ground pins on chip on one of these layers is good practice. Separate ground pins and internal ground wiring for output drivers, for input drivers and the logic core are also helpful.

Static and dynamic characteristics of input/output cells

The same logic provided to test ground noise can be used to test input/output drivers. In contrast to the logic core cells, the output drivers have to drive relatively large capacitive loads and therefore often contribute largely to the total delay of a logic path from input to output pin. It is not normally possible to measure the delay of a single driver, only the combination of input and output driver (2), (3), (1) can be tested. All static input/output characteristics as described in section 4.2 for SSI/MSIs must also be measured for asics. Again, the dynamic behaviour of the tristate and bidirectional drivers including enable and disable times are important.

Internal clock skew

For small asics, ringing and crosstalk on chip is usually not a problem because of the relatively short line length. For newer very large asics containing millions of gates, this will become a problem because of the smaller line distances and multilevel wiring and because of the higher speed. One can expect that the same solutions as were used for board design, namely a single clock system, will be used in chip design too. In a single clock system, the next clock pulse comes when all ringing and crosstalk noise has ceased in the whole chip. The precondition to achieving this is a minimal clock skew between all clock inputs. But large clock nets running over the whole chip, which are characteristic for single clock designs, give considerable clock skew. So special

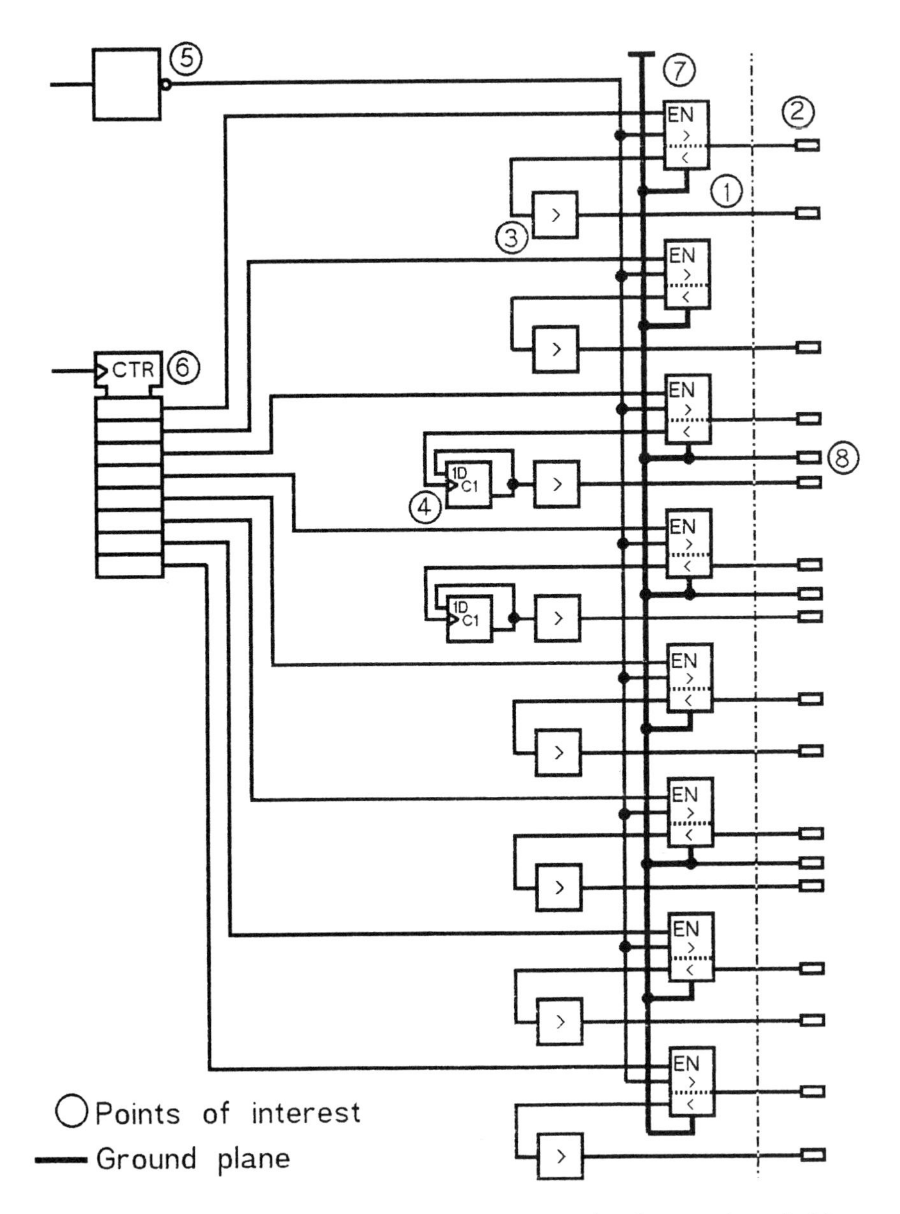

Figure 4.35 Testing ground noise in asics. Many simultaneously switching outputs make it difficult to hold the ground noise of asics under control during test and during operation. The direction of signal flow of bidirectional cells (7) is controlled by the outputs of a counter (6). These cells are driven simultaneously by a gate (5), when they are set in the output state.

measures, like clock distribution lines with equal length or balanced loading, have to be provided. Special software tools are necessary for this. The success of these measures is difficult to check.

Safety tests

Modern asics should be failsafe. Nevertheless, it is worthwhile avoiding unsafe designs, even if they only affect the logic not the safety.

Some libraries contain logic cells which are not allowed to be used without restrictions. Only specific combinations of those cells are allowed; others are strictly forbidden in the user manual, because they can destroy the chip. These arrays can only be used if this condition is strictly controlled by software. Examples include so-called transfer gates that are MOS transistors which connect two logic cells directly as a kind of wired OR. The design must exclude the possibility that outputs with logic high and low state can be connected directly by this transfer gate, as it would produce a short-circuit between supply voltage and ground. But there is always some danger that this can happen unwillingly through a logical or technological failure. So, it is wise to check that, in such a case, only the chip could be destroyed and no further catastrophe like burning printed boards could occur.

Another similar potential danger comes from internal cells with tristate functions, which can be used to realize internal bus connections. If all outputs are in a tristate state condition, then the bus will be floating, the connected receiver is logically undefined and may oscillate or latch-up. Either a design method must be used which prevents such a condition from occuring or special clamping circuits have to be used. Figure 4.36 shows an example of a clamping circuit which keeps the line at a fixed potential if the drivers are tristate.

Functionality and effectiveness of software tools

When designing a gate array, several software tools are used for simulation, routeing, etc. Some of these tools are vendor-specific; some are third-party tools. Although these tools are contained in a design kit, specified and placed at customer's disposal by the vendor, they have to run at the user's site, so incompatibility problems may arise. The platform for these tools should be standard workstations; special hardware requirements are less recommendable. It is important for the quality group to check that working interfaces

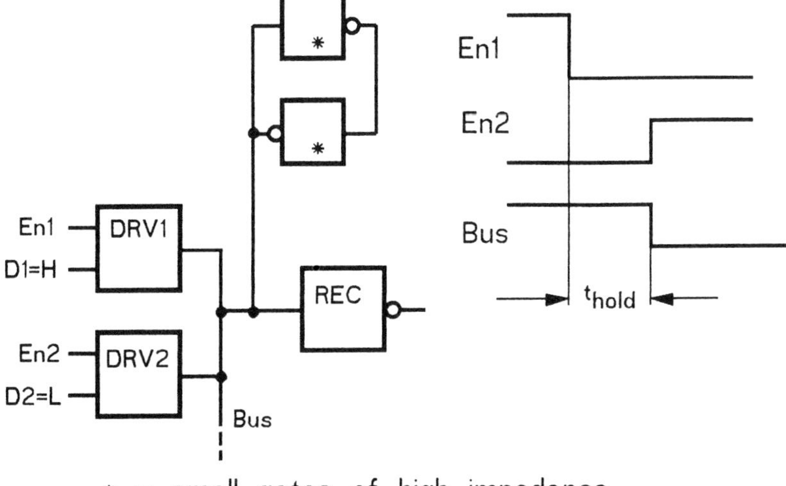

Figure 4.36 Using internal tristate inside asics. Tristate bus lines inside a gate array are sometimes problematic. This hold circuit keeps the bus line in defined conditions.

between the different design tools are provided. The test chip is a good vehicle for revealing these problems.

Testability, effectiveness of test program generators

If the pin count of the gate array exceeds the pin count of the tester, then a dual path or a three-path test has to be used (section 6.4). Normally, the test pattern to test the gate array has to be provided by the user in the vendor's format. The completeness of these patterns is important, because all samples which pass this test are defined as good parts and cannot be rejected by the customer. The vendor will and can take no responsibilty for this. Further problems arise from integrated RAM or ROM cells. Methods like RAM isolation and BIST are advisable.

Loadboard performance

Testing gate arrays of high pin count causes performance problems on the loadboard. Parts which work well in application may fail in test because of the long lines on the loadboard. A carefully designed loadboard is necessary. The test chip is also a good vehicle for testing your loadboard and setting up a correlation between the vendor's outgoing test and the customer's incoming inspection.

Mounting on boards

The test chips should also be used to evaluate the correct manufacturing tools for soldering asics with new packages onto boards.

Environmental and life tests

The design engineers will thank the quality engineers for using test chips for this purpose and leaving them the first production samples to make prototype tests.

Summary

From the above list, it becomes clear that a test chip should not be just a small chip with some test structures, but a chip with maximum size and maximum gate count and packed into the final package to give relevant results.

But one of the greatest advantages of designing a test chip with a new vendor is that it can check all interrelations between the vendor and customer mentioned earlier in this chapter. The transfer of information such as wiring lists, simulation patterns, test patterns and simulation results can be very tedious when it is done for the first time. Even the physical data transmission may be troublesome because of the large amount of data. Misunderstandings can occur when problems arise. So, a test chip will pave the way for the subsequent designs.

Second source for gate arrays

The same arguments for second sources which are valid for MSIs and standard circuits are also valid for asics: availability and price.

- The vendor may discontinue production.
- A sudden inbalance on the market may increase lead time.
- A quality break may cause a shortage of products.
- Competition makes it easier to achieve a price reduction.

But the problems are that the number of parts is low and the development costs are extremely high. It is not the quantity but the value of the circuits and the security of delivery which should be of concern for the customer. An honest vendor will understand the justness of this demand and will support the customer in finding a second source. The special design flow for the development of asics opens several ways towards solving these problems. Figures 4.37 and 4.38 show the design flow for gate arrays and the possible interface between the first and second sources to reduce duplication of work.

Option 1 is the classical case of two completely independent suppliers. This means double work and a maximum cost. These costs can be reduced considerably if a 'vendor optimized design' is done, a design which can be adapted easily to different libraries. A good practice for achieving this is the use of a high level language for the design. Also, special netlist conversion programs can be used to convert the netlist for the first source to a format which can be understood by the second source.

In option 2, identical cell libraries and identical application rules are required. The main work of logical design and graphical input is done only once. The simulated wiring list and the simulation pattern are sent to both vendors. This option is offered or announced by several vendors. A problem could be that each vendor might offer additional cells.

Option 3 requires that both vendors use identical software to generate the layout. The symbolic layout in grid raster has to be transferred from the first vendor to the second source. The author knows no real-life example of this option.

A close co-operation between the two vendors is required for options 4 and 5. They have to have an identical process and even use identical tools. The interface will be either the tape to generate masks or the mask set itself. This was announced by some vendors.

Bearing these options in mind, the preferences cited at the beginning of this section seem somewhat contradictory. Both vendors' samples have to be exchangeable, which is best guaranteed by options 4 or 5. On the other hand, the vendors should be independent, so the customer's buying department is free to negotiate price.

Option 2 is a good compromise if the second source uses a similar floor plan to match the delay times. The simulated netlist will be retargeted by mapping the netlist to the new target library. By this method, a vendor may be selected or changed after the design has been completed. This gives a little more freedom to the customer.

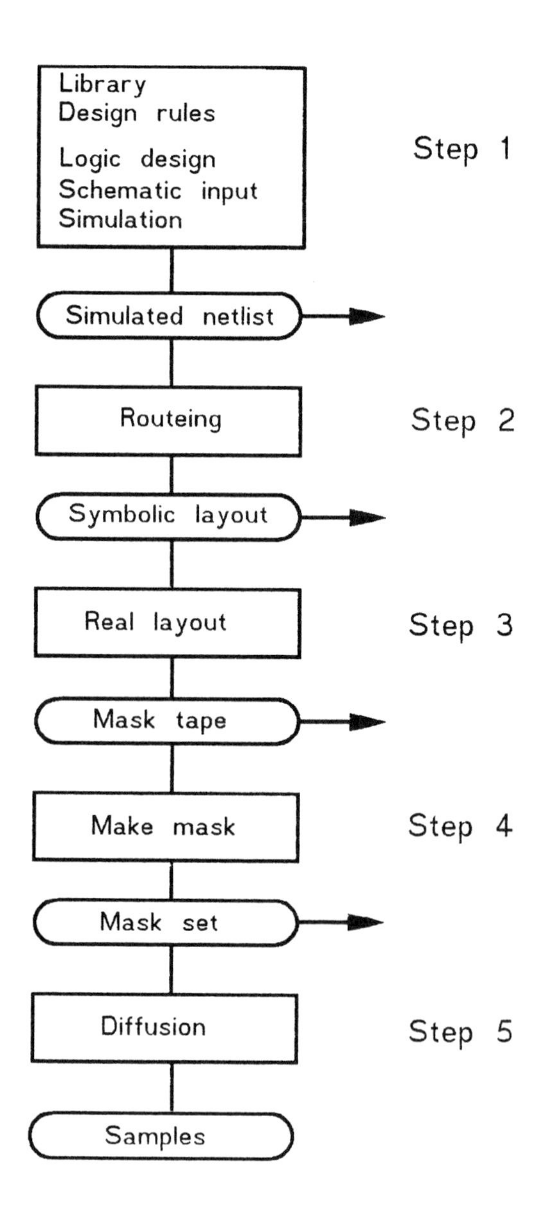

Figure 4.37 Second source for gate arrays. To find an interface to a second source, the design process must be split up.

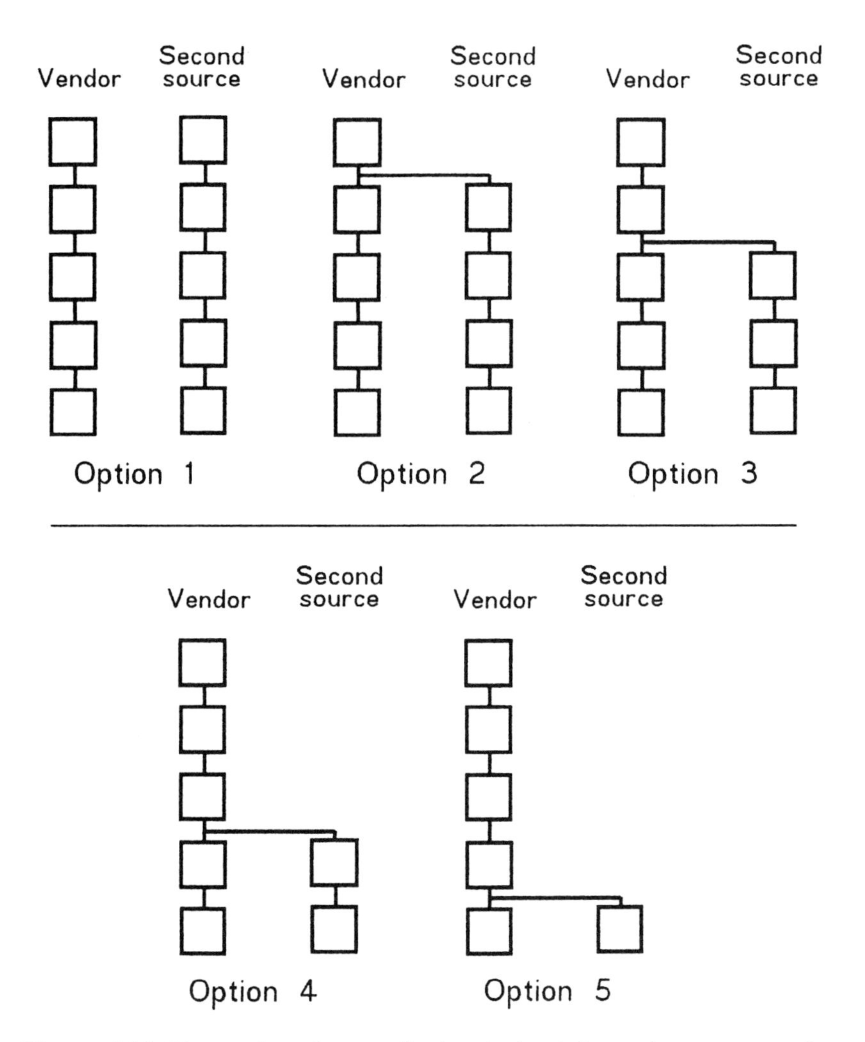

Figure 4.38 Five options for transferring design information to a second source: (1) two independent sources; (2) simulated netlist interface; (3) symbolic lay-out interface; (4) mask tape interface; and (5) mask interface.

4.7.4 User programmable asics

Gate arrays are individualized at the vendor's site. Another class of asics called user programmable or field programmable asics, are individualized by the user. They can be classified according to their logic structure into FPLA, PAL/GAL and PROM/EPROM circuits.

All of them consist of an input matrix, an intermediate active array (to form product terms) and an output matrix. Whereas for PROMs the input matrix is fixed and the output matrix is programmable, for PAL/GAL circuits the inverse is true: the output matrix is fixed and the input matrix programmable (Fig. 4.39). FPLA and FPGA circuits offer the greatest flexibility. The programming was done originally in PROM and PAL circuits by blowing metal fuses. Today, more and more programmable structures are based on antifuses or programmable memory cells like EPROMs. Some PLDs are reprogrammable using EEPROM- or SRAM-based programming schemes. This is particularly attractive for fast prototyping or for applications where the devices have to be reconfigured after being installed in the system. But cheapest and most often used are one-time programmable PAL/GAL circuits with high speed and moderate logic content.

The evaluation of PALs is similar to that of gate arrays. Test circuits have to be programmed in order to measure all critical features. The problem of evaluation is similar too: to find a logic content of the evaluation PAL which allows testing of all critical applications. In most cases, one or two evaluation circuits are sufficient, one circuit for combinatorial logic and possibly one other sample to test register logic. The requirements for that test logic are:

- to test each output;
- to use each input;
- to test at least one path from each input to each output;
- to test the tristate function of all tristate outputs – this is easy if one enable input is provided for all outputs, but more difficult if a product term has to be used;
- to test bidirectional outputs by a sequential feedback from each output to the adjacent one;
- to test for ground noise by simultaneous switching – this is very important for high speed PALs in small packages with a small number of ground pins – all outputs are switched by one input in this test, other inputs should allow to reduce the number of switching outputs one by one, so the maximum admissible number of switching outputs can be derived;
- to test set-up and hold time of buried registers with internally created clock pulses which may be difficult but should not be omitted; and
- to test the response of the registers to asynchronous input signals (problem of metastability).

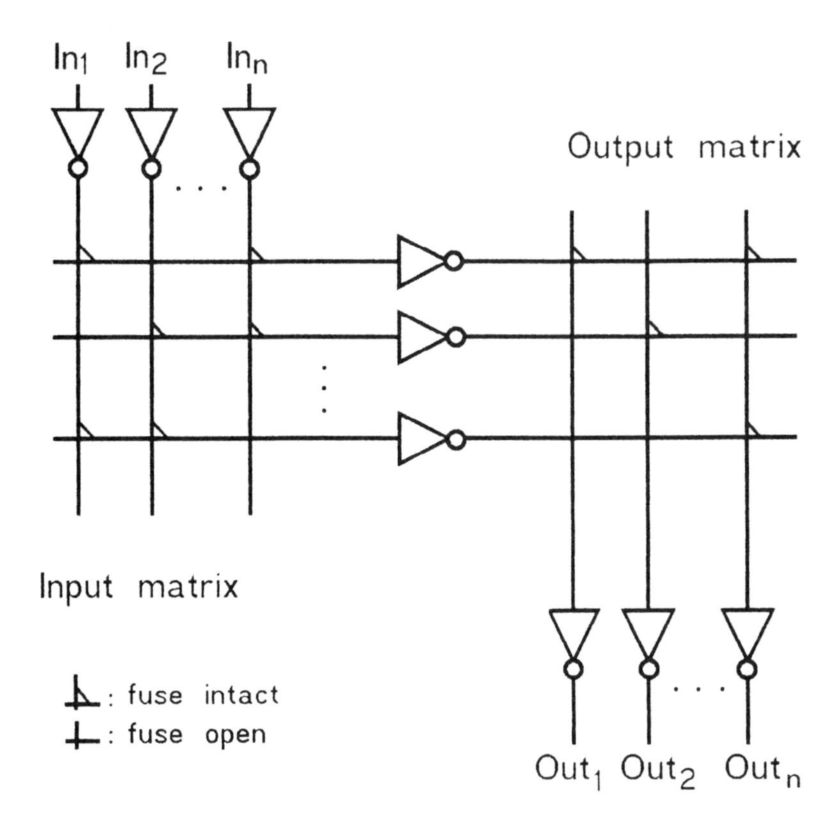

	Input matrix	Output matrix
FPGA	programmable	programmable
ROM	fix	programmable
PAL	programmable	fix

Figure 4.39 Principal organization of user programmable gate arrays.

Although several programs are available commercially which make it easier to generate the test samples automatically, some manual interference seems indispensable in order to allow for structural dependencies. Comparative speed measurements can be made by benchmarks, such as those provided by PREP. Keep in mind that the goal is to detect principal weaknesses of PAL circuits new to the market.

In the manufacturing area, one source of error in the past was a confusion of types during programming. This was because the circuits are handled manually in this production step. Now most vendors provide an electronic signature which eliminates this source of error.

4.8 ELECTRICAL EVALUATION OF MEMORIES

Memories are usually a technology driver, i.e. they are the first components to be made in a new more advanced technology. The reason may be that memories are a mass product with high production volume of relatively few types and with a core of homogeneous structure. There is always a gap of several years until a new generation of memories appear. This is the reason why memories are very intensively evaluated by most customers.

In a first step, initial engineering samples are tested in a similar way to logic products. The function test is much more extensive. Not only the pure logic function but also the influence of internal crosstalk and ground noise are tested. Special test patterns are described in the literature for testing the influence of switching neighbour cells to the cell under test. Therefore, the quality engineer has to know about the geometrical layout of the memory and the assignment of rows and columns to the address pins. This cannot be done by a conventional logic tester. Special memory testers or memory test extensions to ATE testers are used, where the patterns are generated on the fly (i.e. during test). In addition to these systematic tests, tests with randomly generated address patterns should also be made during evaluation to test for non-systematic failures. These tests can remain under full control of the customer (Hayes, 1988).

In a second step, several hundred devices are tested. Distributions of the tested static and dynamic parameters can now be derived. This is the object of joint qualification where the vendor has to deliver detailed data which are checked by the customer's quality group. The third step is a controlled start of production.

All this evaluation and quality assurance work is necessary because of the high volumes of memory components used in an electronic device. The user of memories must not take any risk. This intense evaluation work is the reason memories were among the first components for which ship-to-stock contracts were agreed.

4.9 EVALUATION OF ECL CIRCUITS

4.9.1 Status of ECL technology

ECL circuits have recently lost ground. The probable reason is the rapidly growing integration density of the competing TTL- and CMOS-based circuits. Figure 2.1 illustrates this trend. Besides the higher number of components needed to implement a given logic complex, it is the more complicated manufacturing process and the higher power dissipation of ECL circuits which prevents the improvement in the quality of ECL-based devices to the same extent as has been reached with CMOS. This does not mean that ECL technology is antiquated. It had made considerable progress in the past, by introducing ECLIPS circuits. Furthermore, the trend to more integrated asics was also prevailing with ECL. Today, ECL has the highest speed in a well-established technology. The absence of saturation effects makes the internal and external delay much more calculable. Some kind of multivalued logic called 'series gating' simplifies logic design and is more widely used than similar methods in TTL. It is difficult to predict whether this present retreat of ECL will be final (Miles, 1972).

4.9.2 Evaluation tests on ECL circuits

The same principles of joint qualification presented at the beginning of Chapter 3 are also valid for ECL. Due to the fact that the number of vendors and users of ECL circuits is much smaller, the co-operation between both is much closer. The specification of ECL circuits and their application rules are more detailed than is common for CMOS.

The evaluation tests performed either by the vendor or by the customer are, in general, similar to the tests explained earlier in this chapter. However, there are some differences resulting from the different technology. Whereas for TTL and CMOS the threshold voltage is defined by a number of diode drops and so far is fixed, the logic levels of ECL gates are set by a reference voltage and so far depend on design. The measurement of the logic levels and the threshold voltage and their dependence on temperature and supply voltage plays an important role. Because of the emitter followers, the output and input impedances are critical and have to be controlled to prevent oscillations. Several further special parameters like saturation voltage are unique to ECL. All the parameters considered previously in this chapter, like metastability or test chips for asics, are valid for ECL circuits too.

4.10 EVALUATION OF PASSIVE COMPONENTS, DISCRETE SEMICONDUCTORS AND ANALOGUE ICs

Historically, passive components and electronic tubes were the only components available at the beginning of the electronic age. It was the invention of the transistor which let the discrete semiconductors replace the tubes after the Second World War. Analogue and digital ICs emerged two decades later and began to replace the discrete parts more and more. Today, the process of digitalization is reducing the significance of analogue circuits steadily. This trend seems to be set to continue into the near future. Telecommunications, audiovisual devices and automobile electronics are some examples of the rapid change to digital design. Power electronics is the main domain remaining to discrete semiconductors; interface, converters and controllers are examples of niches where analogue components remain. Passive components play an important role even in modern computers. Whereas the number of ICs was reduced considerably by large-scale integration (LSI), the number of passive components was reduced much less. This concerns the number of component types, the volume of ordered components and also the value of these components. Examples of the use of passive components are found in monitors, power supply modules and peripheral devices.

Often, engineers look upon simple components, like resistors, capacitors or contacts, as neglectable in respect to quality assurance and concentrate their efforts on newly designed or complex ICs. Of course, there are reasons to do this. Modern components are often produced in more refined and less proven technologies and so are more error prone. But also some irrational feelings can draw attention away from the more simple towards the more complex components. Evaluation concerns all kinds of components. Even apparently long-established components are constantly subjected to changes: design changes to improve their properties, or changes of production methods to become competitive. Although all vendors strive to maintain quality while improving their components, a setback in application cannot be excluded. Passive components and discrete semiconductors are not the highlights of development but they were continuously improved. Because they were in the shadow of of the tremendous increase in speed and integration density of digital integrated circuits (one order of magnitude every 4–5 years), they did not always get the attention they deserved. The general trend was:

1. higher integration density through smaller dimensions;
2. improved manufacturability by a SMD packaging; and
3. increased reliability through better materials and production methods.

Here are some examples of changes they did undergo in the recent past:

1. Single resistors are replaced by resistor modules in single inline SMD packages.
   Several oscillators of different frequency are replaced by a single chip PLL oscillator.
   The dimensions of capacitors were decreased by high capacitive ceramics and by improved electrolytics.
2. The use of SMD packages allowed double-sided mounting of components on printed boards.
   Chip-sized packages further increased packaging density and reduced insertion costs.
   Components like inductors and relays followed this trend.
3. Delay lines, formerly made on the basis of LC circuits, were replaced by monolithic ICs on the basis of ramping, which improved reliability from 1000 dpm to 100 dpm.
   Monostables are replaced by start–stop counters.
   The properties of transistors were improved so much that selection into classes of different current amplification or cut-off frequency was no longer necessary, and generally no dynamic measurements are now taken.
4. New challenges are on the way qualifying laser diodes used for optical data transmission on printed boards.

This scenario defines the work of the quality group. Few electrical measurements have to be made. One reason may be that resistors and capacitors belong to the analogue components and their electrical properties are less critical when they are used in a digital environment as terminating resistors or blocking capacitors. The evaluation of passive components concentrates on co-operation with the manufacturing department to develop and evaluate appropriate soldering methods for the new packages and to co-operate closely with the vendors for all reliability tests. Frequent audits are an effective method of guaranteeing good quality. Here are some examples of specific reliability problems.

Tantal capacitors are often used as blocking capacitors to filter power supply noise on printed boards. An accidently applied inverse polarity will often result in these parts catching fire and cause much damage to the board. Remedies include the use of tantal capacitors with an integrated fuse, the use of ceramic blocking capacitors wherever possible, and automatic polarity control when inserting the parts by a pick-and-place machine.

Another problem can arise when using several transistors in parallel in order to allow a higher current. The abolition of static and dynamic testing by the vendor forces the quality group to install some selection equipment and to define the test parameters.

4.11 SUBCONTRACTORS (OEM)

There are two kind of subcontractors: OEM (original equipment manufacturers) from whom complete sub-units like power supplies or disc drives are bought, and manufacturers who produce customer-designed sub-units as an extended workbench. Quality assurance applies to both. OEM parts are often treated a little carelessly. Yet they have to be evaluated as rigidly as other components by the same joint qualification procedure shown in Fig. 3.1. It is important that the quality management is consulted early when a new OEM is selected. Close co-operation with the vendor's design group and quality audits at the manufacturing site are the first steps for the quality engineer. He has to convince himself that the OEM has the same quality status as his own company. Low-priced devices from vendors of unknown quality standards may give rise to problems later on and to wearisome redesigns.

The design may influence quality too. In the case of an OEM power supply, for instance, the trend is towards the introduction of wait states and delivery of power on request. Low power consumption and efficient cooling leads to a robust and reliable product and contributes to general power saving.

4.12 FINAL REMARKS ON ELECTRICAL EVALUATION

This chapter should not be a regarded as a recipe of how to evaluate a certain family of integrated circuits but rather a demonstration of the principles of evaluation by examples. New technologies coming to market in the future will require other kinds of evaluation tests. The principles, however, will remain the same.

- Remember that reliability evaluation is only one part of qualification, electrical evaluation is just as important.
- Get as much data from the manufacturer of components as possible.
- Verify and complete these data by your own measurements, remembering that functions of parameters are more meaningful than discrete values.
- Make simulations whenever possible (e.g. by simulation program spice).
- Compare and analyse all data carefully. If you detect any dubiety then ask your vendor persistently for an explanation.
- Use automatic testers and software assistance for evaluation.

You have to make sure that there is no hidden quality risk in new circuits.

5

Reliability and environmental requirements

5.1 RELIABILITY EVALUATION

The prediction of the reliability (failure rate) of electronic components is of great importance in designing electronic systems. The failure rate is affected by the inherent reliability of the components and by the operational and on-site environmental conditions.

There are several methods in use to predict failure rates. These can be found in the literature, and are not covered by this book. Not all these methods are purely theoretical. The parameters used in the prediction equations are derived from actual field data. The measurement of these parameters is the subject of the reliability evaluation.

The same principles that govern the electrical evaluation hold true for the reliability evaluation. Joint qualification, as mentioned at the beginning of Chapter 3 and presented in Fig. 3.1 (p. 25), is beneficial to both parties, vendor and customer. This is because the goal of joint qualification is to reduce:

- the time consumption and costs for the reliability qualification (some tests last 2000h, i.e. 3 months); and
- the number of test samples needed (important for VLSI circuits).

Joint qualification has to be performed for any new family of components before delivery, as well as for qualified families in the case of major changes. This joint qualification is based upon strong contacts between vendor and the system manufacturer (customer) during which:

- all requirements concerning reliability goals and reliability test methods have to be correlated; and
- all information about reliability test results has to be exchanged.

Along with the procedure presented in Fig. 3.1 (p. 25), a technology questionnaire is submitted (Appendix D) by the customer to the vendor of the components. Questions are listed concerning:

- device technology of chip and package;
- reliability test methods, burn-in procedure and ESD prevention; and
- reliability monitoring.

Based on the answers to this technology questionnaire, the correlation between the reliability target, the measures and test results at the vendor's site and the requirements of the customer, decides whether the component family under consideration has passed the reliability qualification or not.

If the test results did not meet the requirements then a reliability meeting will be arranged during which additional tests at the vendor's or customer's site will be fixed. These additional tests are not restricted to standardized tests, they can also be reliability related, i.e. specified by the customer.

In parallel with the steps listed above, an audit at the vendor's site can convince the customer of the completeness and effectiveness of the quality assurance measures and build up trust between both parties. How to prepare, perform and evaluate an audit will be explained in section 5.10.

A quality specification is agreed by the customer and vendor. This specification establishes general requirements for purchasing, delivery and shipment of all components and defines the qualification requirements. As such, it can serve as a guide during the approval procedure.

After discussions on the test results, audits and specification, a final decision is made about approval.

It should be emphasized again that the customer, and not the vendor of components, is responsible for the quality of his end product, for instance an electronic system. Therefore the customer has to gain and maintain the knowledge and ability to judge and, in case of any doubt, to verify the vendor's data. To save costs, he can use the help of a commercial test centre which will perform all cost-intensive tests and technological analysis work. But the final decision has to be made by the customer.

5.2 RELIABILITY TESTS

In sections 5.2–5.8 information is given on how to perform reliability tests and, more importantly, how to judge the vendor's results in the same manner as was done for the electrical evaluation in Chapter 4. Many tests, well standardized by international standards like MIL-STD-833 or CECC 90 000/90 100, are merely mentioned here, while some more interesting reliability problems are handled explicitly.

The reliability tests are an attempt within a limited testing time to simulate the stress to which an electronic component is subjected during actual use. Test conditions more severe than during actual use are imposed on the component to compensate for this lack of time (Nelson, 1990).

A method generally used to measure the reliability of an electronic component is an **accelerated life test**. A rise in the operating temperature (and sometimes the supply voltage) is used to reduce the time to failure. This higher temperature accelerates chemical reactions according to the well-known Arrhenius equation. This leads to the **high-temperature life test** with dynamic operation, considered one of the most important reliability tests. It can be used to calculate the basic failure rate as shown below.

The conditions generally chosen are T_A = 125°C, duration 1000–2000 h. The following is an example for a sample size of 116 parts and zero failures:

Device hours at 125°C	= 2000 h × 116 parts	= 232 000 h
Device hours at 55°C	= 232 000 × 390	= 90.480×10^6 h
Failure rate λ	= $0.92/90.480 \times 10^6$	= 10.2 fit

An activation energy of 0.96 eV and an upper confidence level of 60% was used in this example (Appendix C).

Another important reliability test for plastic packages is the **humidity test**. It corresponds to the leakage test for hermetically sealed packages. The purpose of this test is to simulate corrosion of the internal wiring. There are three methods which are generally used:

1. THB (temperature–humidity bias) test; the classical humidity test. The test parameters are T_A = 85°C, 85% RH (relative humidity), duration 1000–2000 h, minimum power dissipation, reverse bias.
2. HAST (highly accelerated stress test) which decreases the test time further. The test parameters are T_A = 130°C, 85% RH,

duration 72–200 h. The equivalence of the HAST test to the THB test has to be proved beforehand by correlation.

3. The relatively inexpensive pressure cooker test, mostly used to get a quick result.

The importance of these humidity tests depends strongly on the operating conditions under which the parts are used. For use in office systems with a regulated climate, these tests are of less importance than for applications where devices might be stored for a long time under tropical conditions.

For the VLSI circuit with its large dimensions, temperature cycling, temperature shock and mechanical vibrations are of more importance. A description of these and other tests is given as a specification in Appendix I.

5.3 ESD SENSITIVITY

5.3.1 Physical basis

The abbreviation ESD is used both for electrostatic discharge and for electrostatic sensitive device. Electrostatic discharge is a special case of the more general problem of electrical overstress (EOS). It is any flow of current caused by electrical fields across the device pins or internal elements. These fields are initiated through electrostatic charges. These charges can be generated by charge separation at different materials or by electrostatic induction from an electric field. When two materials are in contact, the electrons at the interface are attracted to the material with the higher electron affinity. After separating the materials, the charge also remains separated if the electrical conductivity of the materials is sufficiently low. The amount of charge generation depends on the electron affinities, the surface conditions, the kinetics, the series resistance to earth and the conductivity of the air. Electrical charges can also be induced by an electric field. Some typical values of charge-up voltages of persons are given in Table 6.9.

Electronic components can be destroyed by electrostatic discharge by the following events:

- discharge of a charged object or person through a component to earth – this failure mode is described by the **human body model** (HBM);

- discharge of a charged component through one connector to earth – this failure mode is called the **charged device model** (CDM); and
- if an unearthed component is inserted into an electric field, internal potential differences are induced, which exceed the breakdown strength of the components' dielectrics. This degradation process is described by the **field-induced model**.

The above mentioned discharge events result in damage like:

- metallization melt;
- thermal breakdown of a pn junction; and
- dielectric breakdown of gate oxide.

So, as well as immediate functional faults, parametric failures can result, which again lead to lifetime problems.

5.3.2 Measurement methods

There are two standardized measurement methods in use, which are based on the HBM and the CDM. They represent different failure models; therefore, they cannot be seen as alternatives.

The HBM is most commonly used to test the sensitivity of a component to electrostatic discharge. This is because charging of a component or inserting a component into a electric field can be controlled better than the discharge of a human being. For the HBM an RC discharge simulates the discharge mechanism as shown in Fig. 5.1(a). A capacitance C is loaded to the stress voltage U and then discharged over a series resistor R between any two pins. So $n!$ measurements have to be taken for a package with n pins. Recommended values for test circuit are: $R = 1-1.5\,\mathrm{k\Omega}$, $C = 100-200\,\mathrm{pF}$. From this comes:

Time constant $\tau = (R + R_i)C = 100-300\,\mathrm{ns}$
Energy $E = \frac{1}{2}(C \times V^2) = 0.05-0.1\,\mathrm{mWs}$ (for V=1000 V)
Power $\approx E/5\pi = 0.03-0.2\,\mathrm{kW}$

The test based on the CDM method is shown in Fig. 5.1(b). The DUT is charged through a high-valued resistor and discharged to earth. The realization of this test set-up is usually more critical. Assuming an internal capacitance of 5 pF and an internal resistance of 25 Ω, the following calculations can be made:

(a)

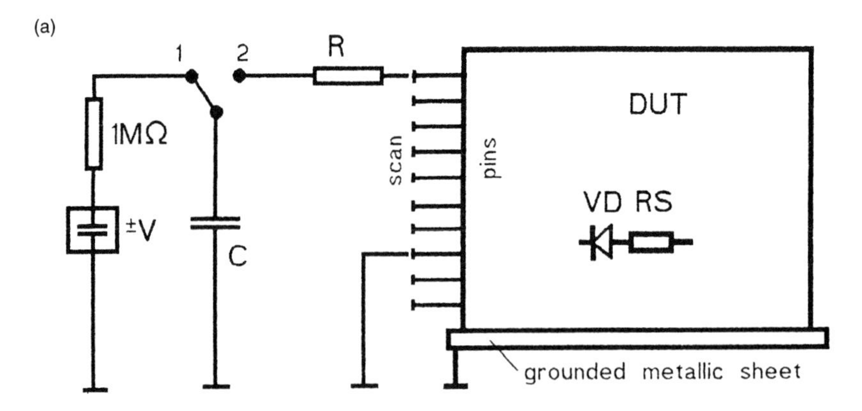

(b)

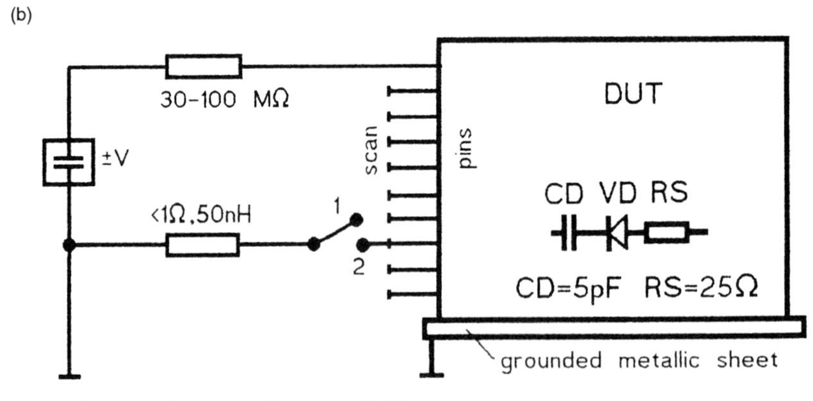

Figure 5.1 Two commonly used circuits to test the ESD sensitivity of a component: (a) human body model (HBM); (b) charged device model (CDM).

$$\begin{aligned} &\text{Time constant } \tau = 125\,\text{ps} \\ &\text{Energy} = 2.5\,\mu\text{Ws} \\ &\text{Power} = 4\,\text{kW} \end{aligned}$$

The above calculations show that, using the CDM, less energy is dissipated in a shorter time which results in higher power.

Each sample has to be tested before and after each electrostatic discharge stress for functional failures and parametric defects. The most sensitive parameter for ICs is the input current which is to be determined by the input characteristic as described in section 4.2.1.

The significance of ESD measurements results from the scaling down of the structural dimensions of modern electronic components, especially ICs. This leads to smaller interconnection width, shallower junctions and decreasing insulator thickness. Whereas the dielectric strength and the thermal capacitance cannot be increased substantially, the sensitivity of a component can be reduced considerably by an appropriate design. So the vendors of modern high-speed components tend to specify ESD sensitivity in their datasheets. In order to benefit from this, the customer should request the measurement data upon which this guarantee is based. As for the test methods the vendors use, refer to standards like MIL-STD-883 Method 3015 or CECC 90 100 Method 4.1.9. Nevertheless, the customer must ensure that measurements are made not only pin to earth but also pin to pin.

The testing strategy for an evaluation test using the HBM is as follows. First, the most sensitive pin combination has to be determined. This means the ESD sensitivity for all pin combinations and for about 5–10 samples has to be determined. Second, a greater sample size for this most sensitive combination has to be tested to get a cumulative distribution N(U) of the stress voltage U (Fig. 5.2). The ESD sensitivity voltage U_{ESD} can be derived from this distribution as the voltage up to which no more than 0.1% of the components fail. Assuming a normal Gaussian distribution, you can derive this voltage without testing thousands of samples.

5.3.3 Results

The following classification of components seems practical:

Class 1 $U_{ESD} < 500\,V$: the component is extremely sensitive and a qualification is not recommended.

Class 2 $500V < U_{ESD} < 4000\,V$: the component is sensitive to electric discharge. It is acceptable if proper ESD handling is realized.

Class 3 $U_{ESD} > 4000\,V$: the component is not sensitive to ESD.

A successful strategy for preventing electrostatic discharge affecting the components in the factory is based on three steps:

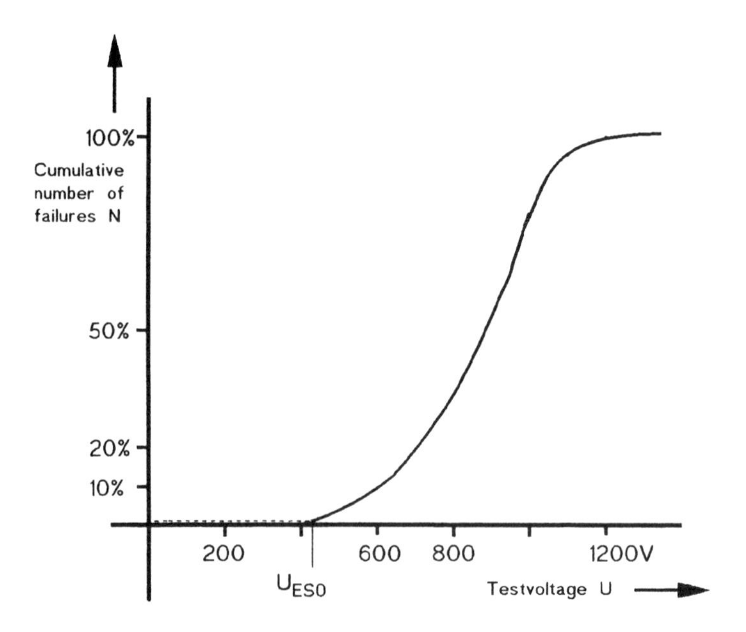

Figure 5.2 How to derive ESD sensitivity from measurements.

1. test the sensitivity of all components to electrostatic discharge;
2. decide on the acceptance of a component from a sensitivity classification; and
3. install proper ESD handling in the factory if necessary.

In most higher-speed electronic devices, the use of components of Class 2 is inevitable; this is particularly valid for ICs. Therefore, measures of protection against electrostatic discharge have to be taken in this case. Some of these measures are:

- installation of protected working areas with dissipative surfaces on all objects;
- earthing of all objects and persons (by a bracelet); and
- safe transport of components through unprotected areas in packages of volume-conducting materials.

These measures are discussed in detail in Chapter 6.

To answer the question of how many of these precautions are really necessary, a quantitative correlation between ESD sensitivity and ESD handling has to be established. Figure 5.3 shows the measured

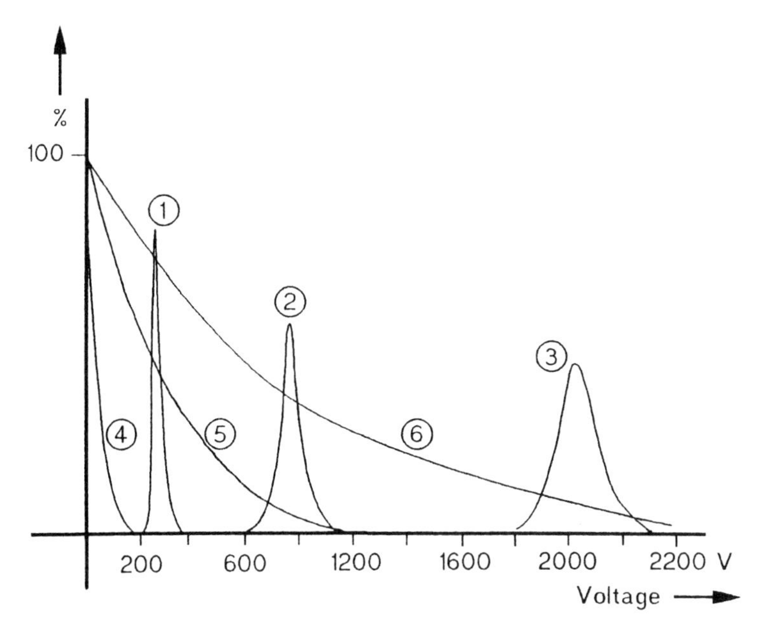

Figure 5.3 ESD sensitivity and ESD potential, the keys to calculating ESD related failure rate. 1 = sensitivity of ECL without protection; 2 = sensitivity of ECL and former CMOS; 3 = sensitivity of modern TTL and CMOS; 4 = potential distribution with ESD handling; 5 = potential distribution with moderate ESD handling; 6 = potential distribution without ESD handling.

sensitivity distribution $n(U) = \mathrm{d}N(U)/\mathrm{d}U$ of several typical components plotted against the 'potential distribution' $\Phi(u)$.

The latter means the probability of a component coming into contact with an electrostatic discharge voltage U during handling. Figure 5.3 is only an example, but if you establish a similar figure for all components in your device and for your own manufacturing area, then you can save considerable expense. It makes a big difference whether your personnel need protective clothes and shoes or only conductive bracelets. But to do this without risk, it is important to remember to:

- evaluate the ESD sensitivity of all component types you use – the weakest component defines the necessary precautions;
- come to an agreement with the vendors to guarantee ESD sensitivity and to do regular re-evaluations; and
- lay down these agreements in a specification (Chapter 6).

5.4 LATCH-UP EFFECT

CMOS circuits have intrinsically parasitic built-in bipolar transistors which may form thyristor circuits (Fig. 5.4). When these thyristors are triggered by external conditions, a latch-up phenomenon can occur. That means the thyristors will stay conducting after the trigger conditions are removed. An excessive current is flowing from V_{CC} to earth which is detrimental to device operation and reliability and results in a catastrophic failure. To minimize the problem the circuit designers lowered the current gain of the transistors and added input protection circuits. Today, no latch-up should occur in normal field use with properly designed CMOS circuits, even if they are new designs with fine structures. On the other hand, the constant miniaturization of CMOS circuits will exaggerate the latch-up problem. The most frequent reason for latch-up is a noise current

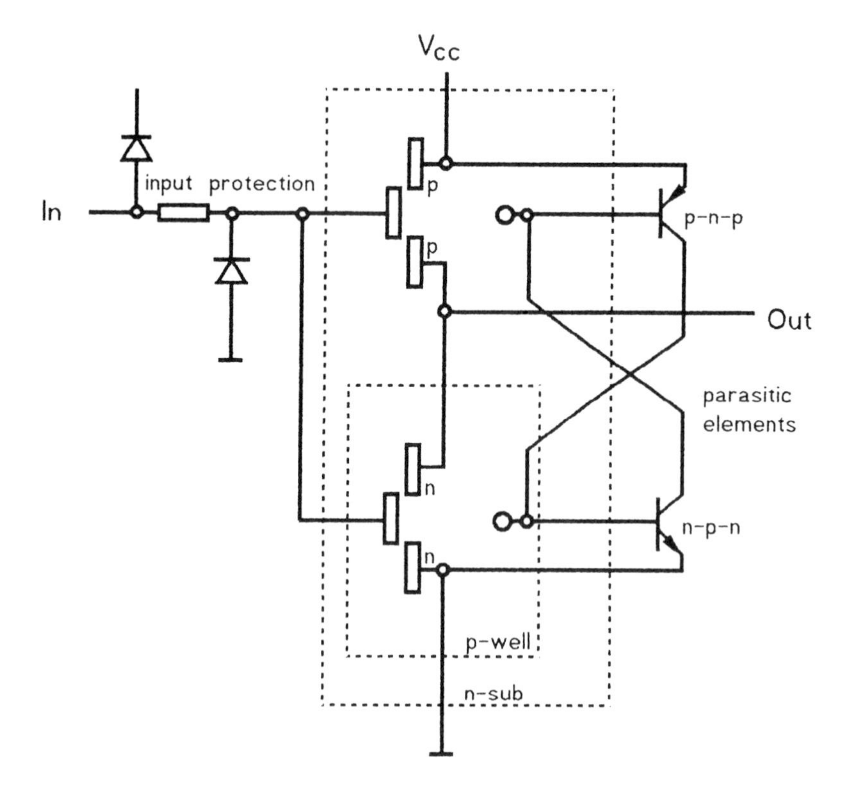

Figure 5.4 Parasitic elements in a CMOS circuit may cause latch-up.

flowing into or out of a pin. For example, this may happen if a board is plugged into a back plane with the power supply on. Therefore, it is wise to do latch-up tests during the evaluation procedure. Two methods are used by the vendors to measure the latch-up sensitivity of their circuits:

1. **The capacitor discharge method** A capacitor (usually ≈200 pF) is charged to a high voltage and then switched to the pin under test. Charge voltages should be greater than 200 V without showing latch-up.
2. **The constant current method** A constant DC current is forced into the pin under test for a short time (less than 1 s) while the supply current is monitored.

It is better if the circuit under test is operating dynamically while tested. If a latch-up occurs, the supply current stays at an elevated level after the current is switched off. Positive and negative currents of greater than 200 mA should be allowed for a well-designed circuit without latch-up occurring. Do not forget to limit the supply current and the input voltage to prevent damaging the circuit. All pins – input, output and supply – have to be measured. To test the supply pin of a 5 V device, the supply voltage is set to about 8 V for 1 s. It is easy to perform latch-up tests with an ATE test system. If the latch-up sensitivity seems critical, the quality engineer should recommend to the logic designer that some safety measures are included. Series resistors of about 100 Ω on all lines leading from the board connector to a critical input provide one method (Croes and Hendrics, 1965).

5.5 SOFT ERRORS

Soft errors were a major problem with DRAMs years ago. A soft error is a malfunction which is not permanent but recoverable by rewriting. The reason for soft errors are α-particles originated from (in descending order):

- coating material of the chip;
- ceramic package;
- plastic package; and
- cosmic rays.

There are two main failure mechanisms causing soft errors:

- **Cell failures** – an α-particle hits a RAM cell and changes its charge. This change can, of course, happen only in one direction, from H to L. The failure, when it occurs, remains stable until new information is written into the cell.
- **Bit line failures** – an α-particle hits a bit line or a sense amplifier and simulates wrong data; both directions of data change occur with equal probability. This failure is transient, and reading a second time will give the correct data.

In the past, these effects were reduced to a tolerable degree by using proper materials and by careful design of DRAMs. A soft error rate of 10–30 fit has been achieved.

But recently, soft error problems have been gaining more interest because:

- the geometry of chip structures became smaller, the system frequency is increasing, and the supply voltage may change from 5 V to 3 V, all of which increases the sensitivity to soft errors;
- there is more demand for hardened chips from military and space applications; and
- the limit of tolerance for soft errors is decreasing.

For the production of high-quality devices, it is indispensable that the soft error rate of critical components like DRAMs is determined as part of a successful qualification. Measured data has to be supplied by the vendor and the suitability of its test equipment has to be checked during an audit. Performing some own soft error tests may prove very helpful in validating the data. Not just DRAMs, but also other complex ICs like microprocessors and asics, were suspected of being affected by soft errors. It is more difficult to test such circuits than DRAMs because failures are transient, i.e. highly sporadic. Continuously running fast go–no go self tests inside the chip can be used as soft error detectors whereas, for DRAMs, a memory tester is used.

It is no longer necessary to install physical equipment to test the soft error rates of components in the course of the evaluation procedure. Formerly, system evaluation had to be made over a long period of time (several months); now fully automated PC guided test systems are commercially available. The principle of these test systems is to accelerate the test (to days) using artificial radioactive sources (e.g. americium 241). Different calibrated filter foils are inserted into the ray trace to model the energy spectrum of the real package source. Figure 5.5 shows the intensity in pulse count as a function of the energy with different filter foils inserted for a certain test system. In the test system, the uncapsulated DUT is exposed to the source and the soft error rate is measured as a function of the distance between DUT and source.

The resulting curves (Fig. 5.6) are extrapolated to distance zero (intersection with the y-axis). These values, the measured soft error rate as function of the energy, are plotted in Fig. 5.7. The soft error rate of the DUT is finally calculated by integration of this plot using a layer model based on the known amount of radiation of the package into which the DUT will be mounted. All this is performed automatically by the test system (Seichter, 1992).

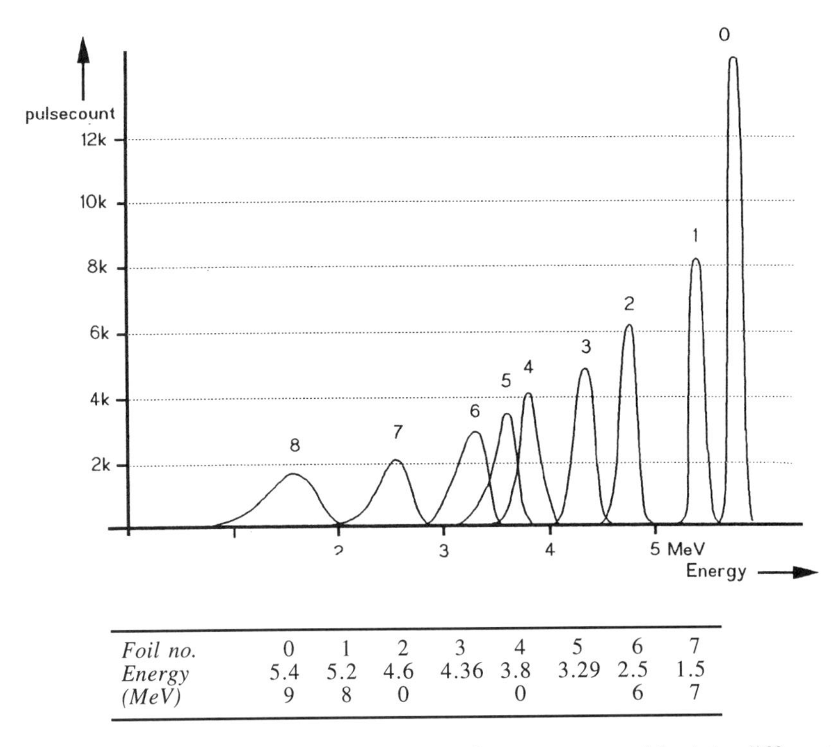

Foil no.	0	1	2	3	4	5	6	7
Energy	5.4	5.2	4.6	4.36	3.8	3.29	2.5	1.5
(MeV)	9	8	0		0		6	7

Figure 5.5 Intensity/energy distribution of an α-source with eight different filter foils.

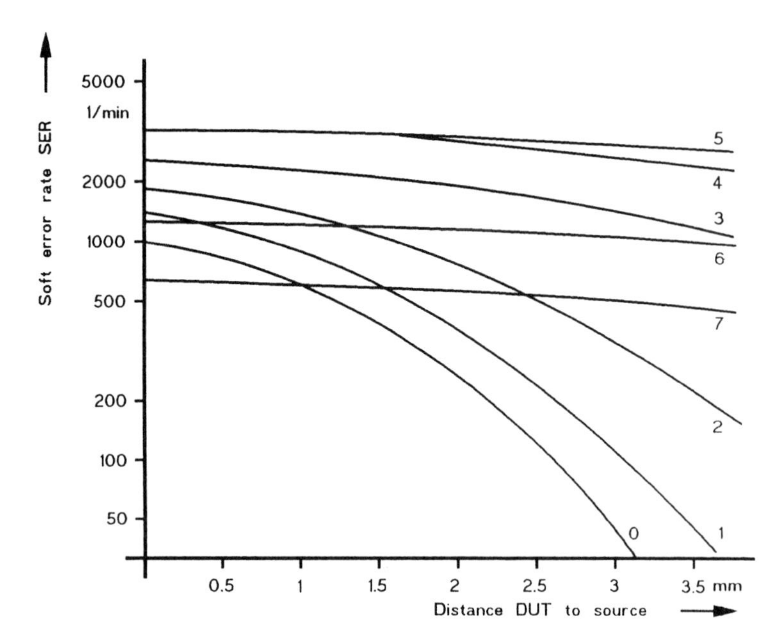

Figure 5.6 Measured soft error rate as a function of the distance between α-source and DUT.

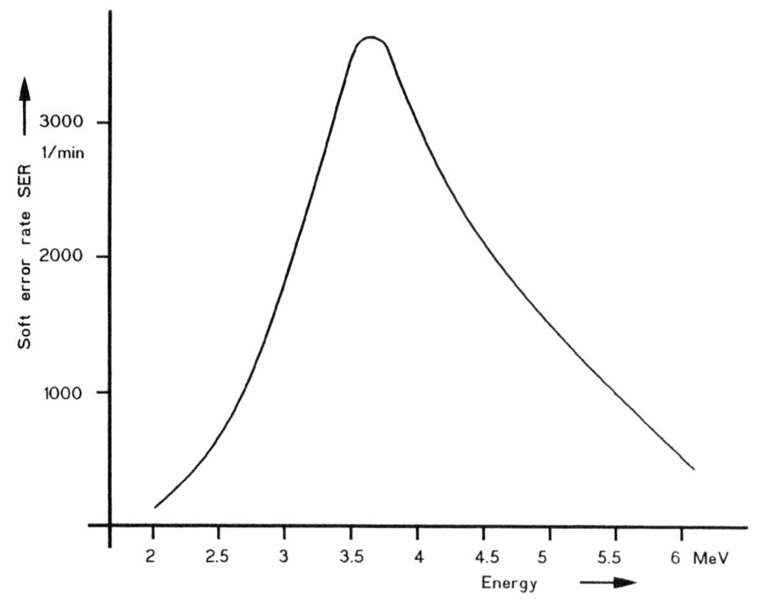

Figure 5.7 Soft error rate as a function of energy.

5.6 PRECONDITIONING (BURN-IN)

Preconditioning means treatment of components before assembly, to screen out early failures. The question for the user is whether it is worthwhile taking preconditioning measures. The usual method for preconditioning beside high-temperature storage is burn-in. The failure rate of an electronic device as a function of its lifetime is known as the bath tub curve (Fig. 5.8). At the beginning the failure rate λ is high (infant mortality) and then decreases to a constant value, the basic failure rate. The physical reasons for infant mortality are oxide failures and mobile ions; wear-out is mostly caused by electromigration (section 5.7). The goal of preconditioning is to prevent infant mortality from affecting the costs of warranty repairs. The main method for achieving this is to simulate early operating life by burn-in and so screen out early failures (Flaherty, 1993d; Pantic, 1986; Romanchik, 1992a; Seichter, 1992).

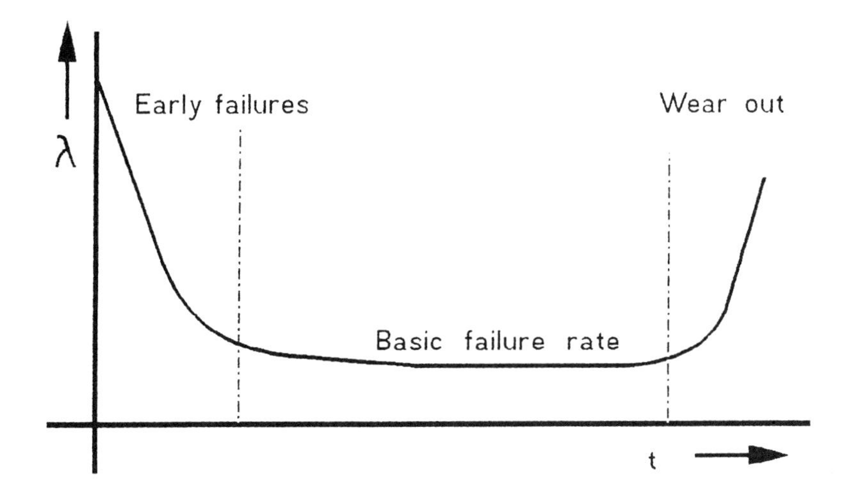

Figure 5.8 The bath tub curve. The failure rate of a component as a function of time shows infant mortality and wearout.

To decide whether burn-in by the user is necessary and will save later warranty costs, a knowledge of the failure rate as a function of time is necessary. Today the quality of circuits should be so much improved – by better production and by an effective burn-in at vendor's site – that no burn-in by the user should be necessary. But this has to be proven by the vendor. The cost of a 100% burn-in for all components is considerable and therefore the vendors are inclined to reduce burn-in and to do only as much as they think absolutely necessary. RAMs, microprocessors and asics are usually burned in by the vendors because early failures of these components will affect the quality of their customers' devices in a conspicuous way. For other components, the vendors try to reduce or omit burn-in. It is advisable to ask the vendor in detail about burn-in conditions, sample size and test results in the course of a joint evaluation procedure. Only if the vendor cannot supply relevant data should users need to take measurements. To measure the failure rate by a monitored burn-in is rather expensive for the user. To save costs, it is best done in co-operation with the vendor.

The user will perform the burn-in and the failure analysis in co-operation with the vendor and share costs. This will give the vendor a good chance of improving the quality of the circuits through corrective actions and getting immediate feedback. About 10^5 samples have to be burned in and tested every 10 hours to obtain relevant data from today's components, as shown in Fig. 5.9. Therefore this can only be done once for each vendor and furthermore only with cheap components like SSI circuits or RAMs. Figure 5.10 shows an example of a measured curve of failures $Nf(t)$ for a total number of Nt = 38 000 small-scale ICs of different types and vendors. Only burn-in related failures were considered. By the formula:

$$Nf(t) = Nt \int_0^t \lambda \, dt$$

the values $\lambda(t)$ can be derived by differentiating the obtained cumulative failure rate of Fig. 5.10. This gives an early failure rate of 136 fit for the first 3000 hours and a basic failure rate of 16 fit after 5000 hours for this example. Modern components may have considerably better values. A recommendation for the valuation of infant mortality for the first 3000 hours is:

less than 50 fit: good quality
more than 250 fit: bad quality.

The burn-in time accelerating factor F_t, which is defined as the relation of the simulated lifetime at operating conditions to the test time at burn-in conditions, is derived from the 'Arrhenius equation':

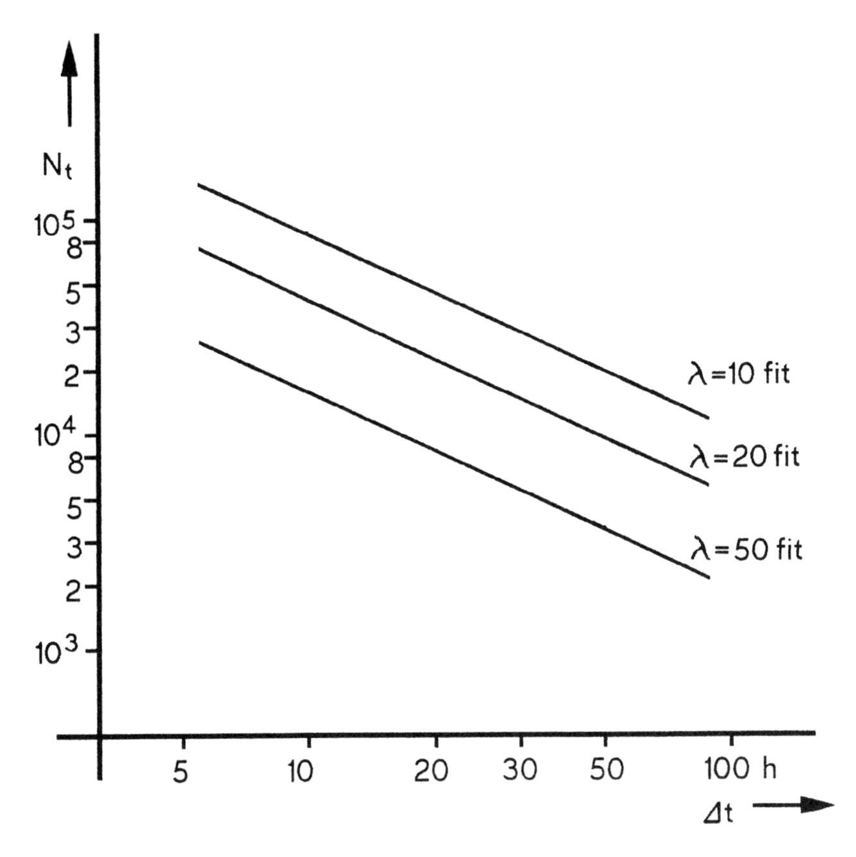

Figure 5.9 A great number of parts is needed to control the early failure rate. N_t is the required sample size and Δt is the smallest time interval between measurements.

$$F_\lambda = \lambda_b/\lambda_0 = \exp\{\Delta E/k(1/T_o - 1/T_b)\}$$

where:

λ	= acceleration factor
k	= 'Boltzmann constant' 8.62×10^{-5} eV/K
T_b	= burn-in temperature
T_o	= operating temperature 333 K
λ_b	= failure rate at burn-in
λ_0	= failure rate during operation
ΔE	= activation energy 0.3 – 1.0 eV (0.3 eV MOS, 0.45 eV bipolar).

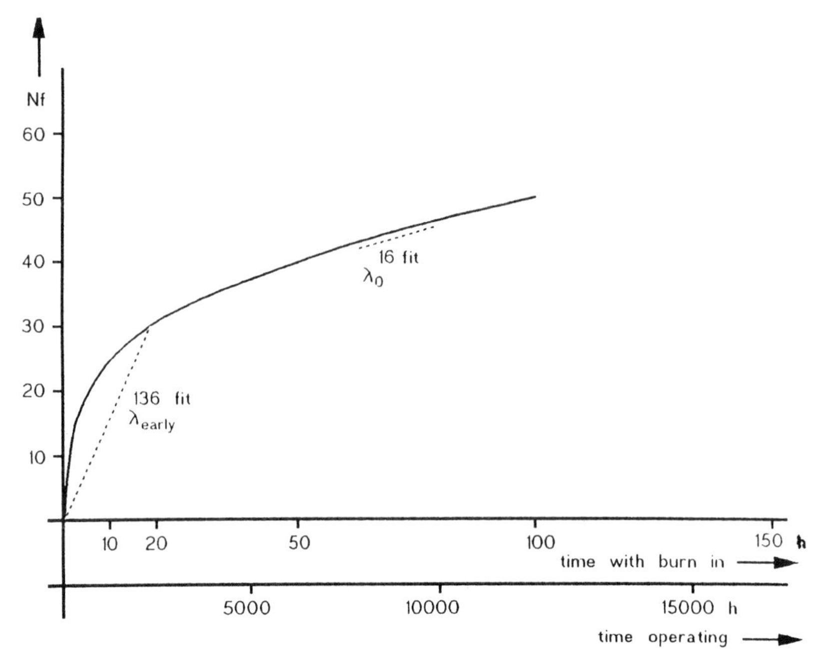

Figure 5.10 The measured failure rate is used to derive the lifetime λ.

Although this equation is valid only for the λ-acceleration factor and only for the time period where the failure rate is constant, i.e. after the infant mortality phase, it can be used as a calculation method to derive the time acceleration factor:

$$F_t = \frac{t_b}{t_o} \approx F_\lambda$$

The time acceleration factor of 120 used in Figs 5.9 and 5.10 was calculated as described above by:

$$T_o = 55\,°C \qquad T_b = 150\,°C \qquad \text{and} \qquad \Delta E = 0.6\,eV$$

A more exact method considers the actual values of failure rates during the infant mortality phase. The time acceleration is calculated from the condition

$$\int_0^{t_b} \lambda_b(t)\,\mathrm{d}t = \int_0^{t_0} \lambda_o(t)\,\mathrm{d}t$$

where:

$\lambda_b(t)$ = failure rate at burn-in conditions
$\lambda_o(t)$ = failure rate at operating conditions
t_b = 'burn-in test' time
t_o = simulated life time at operating conditions.

To determine the costs for burn-in is easy and straightforward, but to calculate the profit is a more complex task. If devices are delivered to the customer (i.e. to the field) with components affected by infant mortality then the maintenance and service cost for one device will be high at first and then decrease slowly. However, because new devices will be continuously delivered to the field, this decrease will not occur for the collective in the field. Let $M(t)$ be the number of devices in the field, B the number of components of one device then field rate α

$$\alpha(t) = B \times \frac{\mathrm{d}M(t)}{\mathrm{d}t}$$

is the number of new components coming into the field. The number of maintenance repairs in the field can then be calculated by:

$$Nf(t) = \int_0^t \left\{ \int_0^t \alpha(t)\,\lambda(t-t')\,\mathrm{d}t' \right\} \mathrm{d}t$$

For calculation, the failure rate can be approximated by:

$$\lambda = \lambda_o \left\{ 1 + a \exp\left(\frac{-t}{b}\right) \right\}$$

where the constants a and b are extracted from the measured curve $\lambda(t)$.

By double integration over $\alpha(t)$ and $\lambda(t)$ the number of maintenance repairs with and without burn-in can be calculated and from this the profit gained from burn-in.

The final goal should be to buy only components with such a low infant mortality that an additional burn-in will bring no profit. But to reach this goal, you first have to be aware of the problem, demand relevant data including measurement results from your vendor and make the necessary calculations. The above example gives hints on how to proceed.

There are other kinds of preconditioning to reduce flaws like defective bonds or bad solder points in the soldering of ICs. Thermal cycling or vibration screening is used for this purpose.

5.7 WEAR-OUT

Wear-out is the rise of the failure rate λ at the end of the life of a component. The main reasons for wear-out are the hot electron effect (hot carrier effect) and electromigration.

The hot electron effect originates in impact ionization in a region of a high electric field near the drain. It shows several kinds of reliability-related problems, for instance a shift of the threshold voltage which slowly deteriorates some dynamic parameters. By using transistors, which are structured to prevent high electric fields, as LDD (lightly doped drain source) MOS, the lifetime is increased to more than 10 years. For well-designed components, this phenomenon became small enough to be ignored. What remained was electromigration.

Electromigration is a phenomenon consisting of a movement of metal atoms under current stress. If a uniform mass movement occurs, problems do not arise. But short and open circuits are caused by electromigration in a boundary region between an area of large mass flow and one of small mass flow. Openings in the interconnection metallization path with bad step coverage are the most important considerations with regard to reliability. Because electromigration is strongly dependent on current density, it can be avoided by an appropriate layout and other skilled production methods.

It is very difficult and time-consuming for a customer to make tests for electromigration on actual components himself, because the activation energy is very low. The manufacturers of components use special critical test structures on chip for routine monitoring. Therefore, it is sensible to ask the vendor for detailed test results and to check them during audits.

Whether the increase of λ in the wear-out phase has really been observed in the past may be questionable, but it should be noted that this phenomenon is still very topical for new circuits with small structures and should be given great consideration by the customer. The danger of wear-out related failures is that it is too late to take any repair action when they occur.

5.8 TECHNOLOGICAL EVALUATION

In order to gain trust in the manufacturing skill of new and unknown vendors, some kind of technological inspection of the workmanship is advisable. The methods and the items of interest are similar to those which will be described later in section 6.9 for failure analysis and are therefore not treated in this chapter.

5.9 MECHANICAL CHARACTERISTICS

The miniaturization of electronic equipment increased the mounting density of electronic components during recent years. The chip size has been scaled up too, due to the high integration. At the same time, the failure rate of components decreased as a result of the efforts of the component manufacturers. About 80% of failures at board test are caused by manufacturing, as will be explained in section 6.5.

Therefore the evaluation of the mechanical properties of electronic components in general and of ICs in surface mountable packages (SMD packages) in particular is gaining significance.

There are two major types of packages: hermetically sealed, which were used for high reliability applications, and plastic encapsulated for normal commercial applications. Today, plastic packages have improved so much that their quality is assumed to be equivalent to hermetically sealed ones. In principle, components delivered in plastic packages need not be inferior to metal–ceramic ones for instance. Nevertheless some key manufacturers prefer metal–ceramic packages – chip carriers (LCC) or pin grid arrays (PGA) – for devices with a high pin count. However, plastic packages as quad flat pack (QFP) with gull wing leads are standard for high pin count (Fig. 5.11) and small outline packages (SOP) for low pin count. New versions of packages will add to this variety in future and they will probably use still finer geometry. Even caseless packaging, i.e. mounting chips directly onto the printed board, is underway for most advanced designs (Flaherty, 1993a; Jones, 1993).

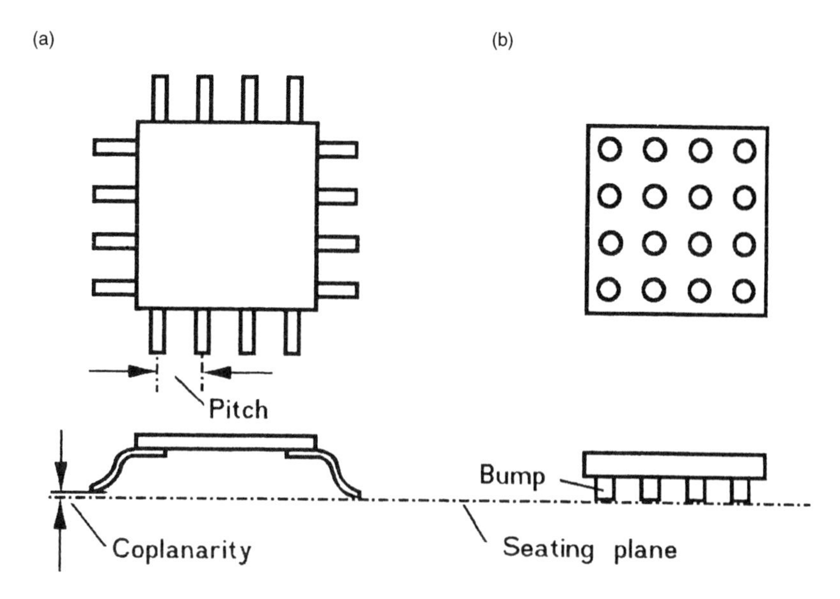

Figure 5.11 Two commonly used packages: (a) quad flat pack with gull wing leads; and (b) plastic pin grid array.

The use of PGA packages (Fig. 5.11(b)) is an alternative to yet more fragile and hard to handle packages. The contacts are small bumps evenly distributed over the lower surface of the package. The advantage is a coarser frame of rigid contacts, less inductivity, better heat transfer and better wireability. However, solder inspection and a resolder process are more difficult. The bumps are almost inaccessible after soldering.

What will the expected quality of an assembled board be when new package types are introduced? That is the question to be answered by the evaluation of packages. This can be done by inspecting the bare package, where the vendor's data are helpful, but it also has to be seen in conjunction with the specific manufacturing process of the user. The qualification engineer has to ask: Is this component suitable to be mounted by a pick-and-place machine? Is it solderable without problems?

The qualification engineer has to demand all available test results from the vendor's assembly site as part of a joint qualification. As well as the package-related tests mentioned earlier (temperature cycle, thermal shock, moisture tests) there are special tests for solderability, lead integrity (lead fatigue, adhesion of lead finish), internal water vapour content and flammability which are performed by the vendor on all new packages and the results of which are made available to the customer. Customers have to evaluate all the data and take some control measurements to ensure the vendor's data is understood correctly.

First, it is necessary to look at physical dimensions, lead material, lead shape and package coplanarity. Coplanarity is the worst case distance between a lead and its target pad on the board. It results from the fact that some leads always fail to reach the seating plane. Solder paste can offset a certain amount of coplanarity and so form a good contact. The paste, when melted, bridges the gap if its height is more than the coplanarity. So the coplanarity is a very important parameter which has a great influence on the quality of the assembled board. Four mils (0.1 mm) is a tolerable value for coplanarity. Due to the higher component density in future it is necessary to use printed boards with reduced line width and pad size and components with a much smaller pitch. This in turn reduces the volume of solder paste and forces a tightening of the coplanarity to guarantee good soldering quality. The shape of the leads compensate for a thermal expansion of the package. There is some advantage in using the gull wing geometry.

Second, the customer has to look at lead surface material and lead coverage. A gold surface is appropriate for socket mounting but not for wave soldering. Silver should be avoided because of corrosion or migration. Tin–lead solder dip is widely used today. Pure tin plating is also used but it is going to be replaced by solder dip because of the danger of tin whiskers. Some vendors use other kinds of surface, material, like palladium. It is up to the users to test the suitability of these surfaces to their own manufacturing process. Insufficient lead coverage is another problem which arises sometimes. If the leads are cut to length and shaped after plating then it is possible that the surface of the cut ends will not be adequately covered. A lead coverage of greater than 95% is usually required.

Future packages with 800 or more leads and with lead pitches of 15 or 20 mils (0.4–0.5 mm) require increased precision of all package dimensions and of the flatness of the seating plane so as to be placed in correct alignment. Higher speed will force the design engineers to consider lead capacitance and lead inductance in spice simulation. So package evaluation will gain importance.

Several additional test procedures have to be carried out by the customers to check the conformance of new packages to their manufacturing process or to check a changed soldering method on existing packages. The conventional wave soldering may have been replaced by a reflow method or vapour phase by infra-red soldering. Package quality, placement and solder process must fit each other. Little help can be expected from the vendor in this respect. To check this, test boards have to be designed, manufactured and evaluated. The general procedure is:

- packages under test are mounted and soldered on test board;
- all solder points are controlled optically;
- stress is indicated by thermal cycling;
- optical control is reapplied; and
- solder quality is tested, e.g. the force necessary to pull off the leads is measured.

In each case, the quality engineer has to co-operate closely with the manufacturing engineer to discuss the test results and decide on the approval of a new package. This can be done using the procedure of FMEA which is explained in Chapter 6.

Equipment to perform lead inspection is commercially available. It is usually based on optical images, but seldom on laser images. X-ray systems have to be used to inspect inaccessible solder points optically.

In the last few years, the so-called cracking phenomenon appeared when soldering large-sized plastic surface-mounted ICs. This cracking is caused by solder heat if the package resin had absorbed moisture from the surrounding atmosphere. Due to the rise in temperature during soldering, the absorbed water content vaporizes rapidly and the vaporized moisture expands the volume, generating cracks. This effect is related to resin thickness and chip size and occurs mainly on large packages of small resin thickness. It is recommended that either a baking procedure is employed to dehumidify the package or better to buy plastic surface-mounted components with more than 24 pins in a moisture-proof packing with desiccants and use them as soon as possible after breaking the seal. This is a problem of concern to the manufacturing management.

Last but not least, it has to be emphasized that the package problems of passive components must not be neglected. As was mentioned in section 4.10, the miniaturization of passive components and the attached change in packaging, though less spectacular, has made constant progress. Optical data processing and optical transmission lines on printed boards will become a new challenge.

5.10 CUSTOMER AUDITS AT VENDOR'S SITE

A well-proven method for establishing greater trust between vendor and customer is the performance of audits. Audits are not sightseeing visits but an effective and cost-saving method for customers to convince themselves of the completeness and the effectiveness of the vendor's quality assurance system. They can check the vendor's ability to produce components of high quality and they can build up trust in the vendor's quality data as a prerequisite for a later ship-to-stock contract.

Therefore, each audit has to be prepared thoroughly. To perform an audit, first an agreement has to be made between vendor and customer as to which parts of the vendor's production facilities shall be visited and in what time frame. The customer has to inform the vendor in writing exactly what needs to be seen and the vendor in turn has to inform the customer what will be shown. Competent and experienced persons from both sides must perform the audit. It presents a bad image if the vendor delegates incompetent people. The customers should prepare themselves thoroughly in advance and make a list of all items to be checked at the vendor's plant. Table 5.1 shows such a checklist.

ISO 9000 registration of the vendor can ease the audit but it will not render it unnecessary. A personal view of the vendor's site will still be required.

5.11 COST AND BENEFITS OF EVALUATION

5.11.1 How to calculate the benefits of evaluation

The purpose of evaluation is to reduce the overall costs of quality assurance by a shift of effort from the production phase to the preproduction phase (Chapter 2). The assumption is that costs for failure removal diminish, the earlier the failure is detected during the production phase (Table 2.2, column 3). Therefore a shift to the preproduction phase, i.e. preventing the usage of components of potentially bad quality, should further reduce the cost. It is prerequisite to a reduction in production tests. In spite of the fact that these assumptions are generally acknowledged, not many cost comparison figures are published. The reason for this restraint may lie in the fact that there is little pressure from management and the efforts necessary to obtain relevant data are not trivial. In addition, publishing such cost comparisons may give an inside view to

Table 5.1 Proposal for an audit check list

1.	General	Location to be visited Product spectrum manufactured in this location Age of factory Total area of factory Process steps performed Number of employees Number of working hours per week
2.	Contamination control	Control performed continously? Contamination class adequate for the process step? wafer processing : class 10 assembly : class 100 testing : class 1000
3.	ESD prevention	Protected working areas with dissipative surface (floor, furniture)? Earthing of all objects and persons by bracelet? Safe transport through unprotected areas?
4.	Quality assurance of raw material	Incoming inspection of all material and components? Vendor audits performed? Certification of vendors?
5.	Process control	SPC control: 100%?
6.	Reliability qualification	All processes and products qualified? Certification system in use?
7.	Reliability monitoring	Life tests? HBT test, HAST, pressure cooker test? Other tests?
8.	Failure analysis	Electrical? Technological? Test lab available?
9.	Preconditioning (burn-in)	Product flow? Conditions (temperature and burn-in time) and calculations?
10.	Training	Instruction of personnel? Training system with certification?
11.	Automatization of production	Automatic transport?
12.	Documentation	

competitors and may therefore be disadvantageous. Competition between different divisions within a company may also be a reason for a certain reluctance.

Again, the goal of evaluation is not primarily to reduce the number of failures but to detect the reasons for failures. The object is not to find bad parts but to find the principal weaknesses of new parts. The problem when making cost comparisons is that you have to compare the costs of an engineering team to carry out the evaluation to the previous costs before the existence of this team. The ideal case would be to compare a company without any evaluation activities with a similar company which does a perfect evaluation. But such a situation can scarcely be found. A way out is to perform a kind of incremental comparison as described below. This is a very tedious task which does not come free, and requires additional effort.

Statistics have to be made over a period of several years. The method is to compare the efforts of the evaluation team with the costs of solving the remaining problems by a problem-solving team. By extrapolation of this curve, you will get the costs of more or less evaluation. The data on which the statistics is based are rather sparse in most cases because they depend on many parameters which are often not known exactly and which vary with time. Therefore, each company has to make its own statistical analysis.

The method described below can only be done with the procedure of failure detection and analysis in mind, which is the subject of section 6.9. All faulty parts have to be analysed and assigned to a certain class of failures. For the following procedure, only failures which can be affected by evaluation are considered, i.e. only those parts or production methods which show a principal weakness. All such failures found on the same type of component and caused by the same reason are counted as one problem. All problems have to be enumerated, listed and documented. This can be done during the regular quality meetings to be described in Chapter 6. In the following subsection (5.11.2), some examples of problems which actually occurred are listed to show what is meant. From these statistics, it is possible to derive the number of problems per year as a rough measure of evaluation-related costs. The number of problems is dependent on the production volume. The more devices produced and the more complex they are, the more problems are likely to occur. The statistics have to be gathered for several years, during which time the production volume of the company may vary. If this is the case then the number of problems has to be normalized to a reference production volume.

There are several ways to apply results given here to different companies. One method is to use the number of bought components normalized by the logarithm of their equivalent gate functions as a measure of the production volume. The severity of the problems has to be taken into consideration too. For the following example, each problem was weighted before addition by a factor dependent on where the problem was detected.

$$P_{\mathrm{SUM}} = P_{\mathrm{I}} \times K_{\mathrm{I}} + P_{\mathrm{B}} \times K_{\mathrm{B}} + P_{\mathrm{E}} \times K_{\mathrm{E}} + P_{\mathrm{F}} \times K_{\mathrm{F}}$$

$$P_{\mathrm{REL}} = \frac{P_{\mathrm{SUM}}}{P_{\mathrm{V}}}$$

where:

P_{SUM} = sum of all problems
P_{I} = number of problems at incoming inspection
K_{I} = weighting factor for incoming inspection
P_{B} = number of problems at board test
K_{B} = weighting factor for board test
P_{E} = number of problems for end test
K_{E} = weighting factor for end test
P_{F} = number of field problems
K_{F} = weighting factor for field problems
P_{REL} = normalized sum of problems
P_{V} = purchase volume (number of parts, normalized).

All these problems listed above have to be solved by quality engineers, assisted when necessary by experts. The time they need to do this has to be listed too. No general figures can be given here because they vary from company to company. Rough estimates are from two engineer–days at the level of incoming inspection for failure analysis and change of test program, up to more than 200 engineer–days to solve severe problems at maintenance or service level. The reader might try to calculate his or her own figures from the examples given in section 5.11.2. The following factors have to be taken into account:

- time for failure analysis and to define the problem;
- time to make corrections and/or design changes respectively production changes; and
- time of production standstill.

It is noticeable that the costs for replacement of defective parts are usually negligible in this calculation. As a by-product of this documentation, the following weighting factors were derived.

$$K_I = 2 \qquad K_B = 5 \qquad K_E = 10 \qquad K_F = 50$$

The main result, however, is the real cost-saving potential for evaluation because most of these problems could have been avoided by a thorough evaluation.

The other side of the coin is the effort spent in evaluation. This can be measured by the number of engineers working on evaluation, relative to the number of newly introduced components and their complexity. A certain lead time has to be considered, i.e. the time after approval until a problematic component shows problems which hinder manufacturing. One very important factor must not be neglected in this calculation; it is not the number of engineers but their effectiveness which counts. This effectiveness is influenced by many conditions. The two main factors are:

1. the usage of ATE test systems will increase the effectiveness by a factor of three (section 4.4); and
2. the general acceptance of the role of evaluation, how much evaluation work has to be done and how decisive the results for all departments, is difficult to express in a number.

An example for one company is shown in Fig. 5.12. All curves in this figure are normalized to make them more usable for the readers. Curve 1 is the normalized sum of problems per 10^6 components per year (only ICs were considered here). Notice the steady decrease due to the benefit of evaluation. Curve 2 is the normalized number of engineers per 1000 qualifications, where the influence of the effectivity factor given by curve 3 was considered.

It can be seen that there is a correlation between the evaluation effort and the number of problems. A greater evaluation effort, i.e. more quality engineers for a certain quantity of new circuits, will reduce the number of problems and so reduce quality costs.

This can be seen very clearly by looking at the lower part of Fig. 5.12. This is an example of one company only, but the trend shown here will be valid for other companies too. Curve 4 shows the real (not normalized) number of engineers working for evaluation, and curve 5 the total number of engineers working for evaluation and

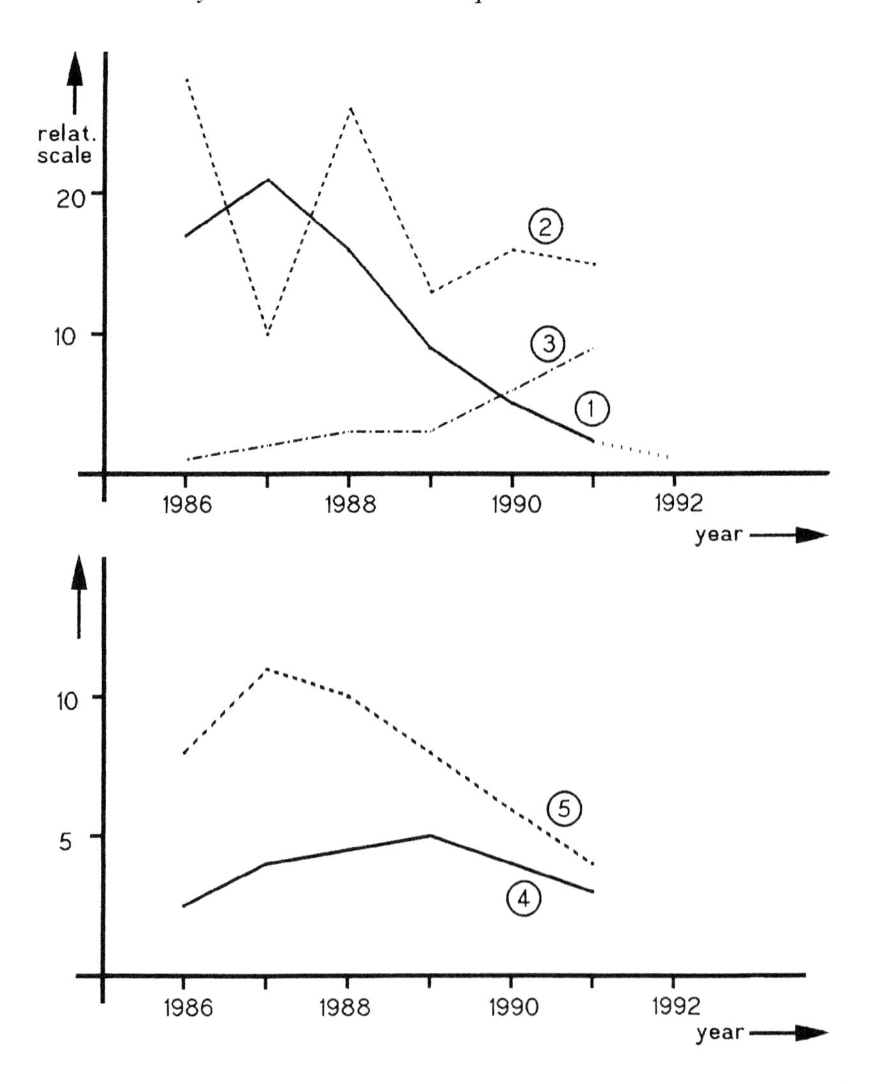

Figure 5.12 Cost and benefit of evaluation. Less trouble at production and service stages indicates that evaluation is worth its costs (see text for explanation).

problem solving. The technicians operating the testers were not included in this calculation, nor was the saving in test equipment.

It has to be admitted that the general improvement in quality, as will be shown in Chapter 6, contributed to this remarkable cost saving too. This improvement, however, depends a great deal on more reliable components, which in turn are the outcome of joint qualification.

5.11.2 Some real problems as examples

In the following list, a few examples based on practical experience at one computer manufacturer are given to illustrate the kind of problems that can occur.

Problem 1: a field problem

In a newly designed mainframe, severe data losses occurred every 2–3 months at the customer's site. The data was reconstructed and suspicious boards exchanged by service engineers several times but without success. Finally, to please the customer, the complete mainframe system was exchanged. A thorough analysis done by a team of experts identified a fifo used to synchronize the data transfer as the cause. Though in the datasheet this fifo's suitability for synchronizing had been advertised, a hidden application note restricted this feature to special conditions. The effort spent in solving this problem involved:

- several boards exchanged and customer's data reconstructed;
- exchange of the complete mainframe system;
- intense search for the reason of the failure including special software to trigger the sporadic failure and store relevant data; and
- redesign of the logic used for asynchronous data transfer.

An evaluation of the metastable behaviour of the fifo would have avoided this.

Problem 2: a field problem

Service people reported a steady increase in failure rate of one component in a computer which had been sold for several years. Analysis showed a wear-out problem (bad step coverage) for this component, for which a vendor change to a second source had been made. All affected boards had to be subjected to a cyclic call-back procedure. A more critical attitude to the reliability data given by the second source and a subsequent audit or a stringent technological analysis would have avoided this.

Problem 3: a problem of end test

During end testing at the lower limit of supply voltage, a high failure rate was found. Analysis showed a critical distribution of the set-up time of a flipflop which would have been detected easily by a dynamic evaluation test.

Problem 4: a board test problem

Severe problems during soldering occurred with asics. Explicit correspondence with the vendor showed that the package had gold-plated pins and was destined for adapter mounting. It was insufficiently solder-dipped by the vendor. About 8 μm of gold was hidden under the tin surface. Joint efforts by both the customer's and vendor's manufacturing departments improved the dipping procedure. A package evaluation including solder tests would have shown up this problem beforehand.

Problem 5: problem at incoming inspection

An asic showed a high failure rate at incoming inspection. No correlation was found with the vendor. The cause was a difference in the failure simulation programs. A closer co-operation between the test groups would have avoided this.

The message of this chapter is: an evaluation is less expensive in time than a catastrophe at production.

6

Quality assurance in the production phase

6.1 GENERAL CONSIDERATIONS

The fast development of technology together with growing competition and shorter innovation cycles caused severe economic problems for the electronic industry. Cheaper products at the expense of consumer quality is by no means a promising response. Past experience shows that improving quality without increasing cost is a better way to compete. Chapter 3 showed how the quality of a product can be increased by better quality assurance in the pre-production phase.

An evaluation of all components prevents many problems and close co-operation with the vendors reduces the cost for this. But the main costs of quality assurance arise in the production phase. Here intensive testing after every production step is the way to raise quality. Because of the considerable effort required, the question arises whether the costs associated with these measures can be reduced. The answer depends on requirements which are specific to the application.

As was shown in Table 1.2 in the introduction to this book, several strategies have been established. Depending on whether products are consumer oriented, military or commercial, more weight is laid on either quality or cost. In each case, however, the final goal of quality assurance in the production phase is to achieve good quality at low cost.

But let me repeat: First, you have to concentrate on good quality, then you can think about reducing cost by rationalized production, proper management and close co-operation with the vendor to minimize testing. The following sections show how this goal can be achieved (O'Connor, 1993; Raheja, 1991).

A quality management system is prerequisite to an effective and successful improvement of the quality of the end product. This is explained in detail in the next section, 6.2. How such a system can be applied to electronic products is shown in the sections which follow.

First, a complete test after each production step is necessary to ascertain the quality of the incoming components and of the production methods. This is the theme of sections 6.3–6.8. All detected faulty components at incoming inspection and at module and final tests have to be analysed to show up weak points. A detailed failure description is necessary (section 6.9). The results have to be correlated with those of the vendor. The next step is consequently an elimination of the analysed failures to improve quality. Good failure statistics will show the improvements. As soon as the quality has improved to a satisfactory degree, the tests can gradually be eliminated until only quality monitoring remains. How all this can be done is the subject of sections 6.10 and 6.11.

6.2 QUALITY MANAGEMENT

6.2.1 Failure management system

The manufacturing process is composed of many single production steps, each followed by test procedures. Quality management is a procedure which covers all steps of production from design, purchasing and incoming inspection, manufacturing and board testing to final testing and maintenance. Its main functions are to produce a survey, a documentation of all failures, to correct these failures and to eliminate the sources which caused them. The object of quality management is to show up quality problems in products, procedures and processes, as well as to recommend, establish and set into effect solutions to solve these problems.

Both requests from customers and internal interests are reasons for installing a rigid quality management system. These include the following:

- Increasing demands from customers for better quality and for complete documentation of all quality assurance measures. This is often a top item on customers' audits, where documentation of failures detected by the manufacturer in the production process and the countermeasures is demanded. It has been performed with great success by most semiconductor manufacturers; other producers (e.g. automotives) are also following this path.
- Tightening up of legal regulations like product liability law.
- A requirement for less time to market.
- A request for a clear internal procedure of quality assurance.

- A desire to encourage co-operation beyond the bounds of departments.
- Documentation of incoming and outgoing quality.
- A desire to reduce incoming inspection costs aiming at ship-to-stock.

The preconditions for an effective quality and failure management are:

- specified manufacturing process (this specification has to be available in written form at all production sites);
- well-defined quality goals (the actual quality level should not be the incidental result of *ad hoc* measures – it has to be well planned from the beginning of the design for each manufacturing step since the quality performance is as important as the electrical performance); and
- a failure catalogue at each test location.

All this seems to be possible only with an adequate computer aided database which provides an effective and time-saving way to input, store, interconnect and retrieve all quality data. The goal of such a system should be:

- transparency of actual failure rate and its trend to all personnel concerned with quality (this will considerably increase motivation to reduce failures);
- standardization of methods to avoid or remove failures;
- transparency of costs for repair and consequently the elimination of failure sources to reduce this cost;
- continuous improvement of all manufacturing processes; and
- avoidance of the same failure happening twice.

The methods of achieving this goal are shown in Fig. 6.1 for one of the consecutive manufacturing steps.

The first item is failure detection. Explicit test directives have to be defined for all test steps, e.g. incoming inspection, board test or final test, but also for all eventual intermediate tests. These directives enable the test engineers or technicians to detect known failures in products, processes and procedures either manually or preferably by automatic test systems. These directives consist of a detailed manual of test instructions.

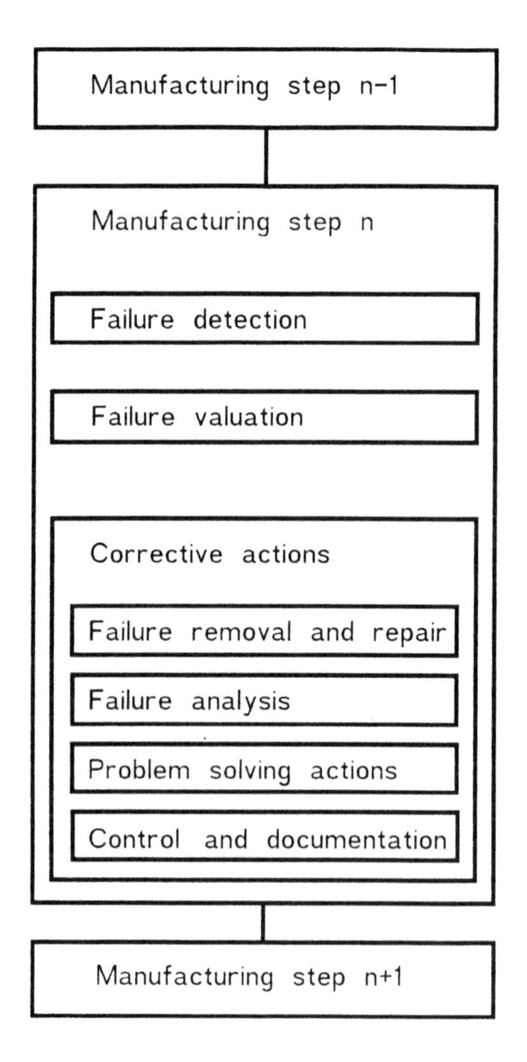

Figure 6.1 Quality management has to control the effectiveness of failure detection and of corrective measures at each manufacturing step.

The second item is a valuation or appraisal of the detected failures. That means deciding to which category the failure belongs and how to proceed.

This leads to the third item which is corrective and preventive actions.

1. Repair or replace the faulty part and repeat the test. If the part is unrepairable, discard it.
2. Analyse the failure to find out the exact reason for it. This may consist of a technological analysis of the faulty component or an analysis of the manufacturing process to check potential physical or organization weakness. Each observed symptom is assigned to a specific failure source in the failure catalogue, preferably with the assistance of an expert system.
3. Solve the detected problems by a series of quality actions decided upon by a quality team:

   - repair the actual failure;
   - prevent its repetition; and
   - correct faulty product designs which may have passed the test with undetected failures.

4. Archive and evaluate statistically all corrective actions. Document with respect to

   - product, process or procedure concerned;
   - failure categories or classes; and
   - expenses and costs of repair.

   The evaluation of these data by the failure management is related to:

   - observing and controlling long-term quality limits;
   - determination of trends;
   - recognition of clusters with regard to reasons and symptoms of failures; and
   - accumulation of costs.

6.2.2 Manufacturing process control

All tests in the course of the production flow, which are executed on the responsibility of the manufacturing department, should secure the continuously running production flow with an output of high quality. To reach this goal they must be assisted by two attributive controlling systems as the responsibilty of quality assurance, product control and manufacturing process control (Fig. 6.2). These have to be organizationally independent from manufacturing.

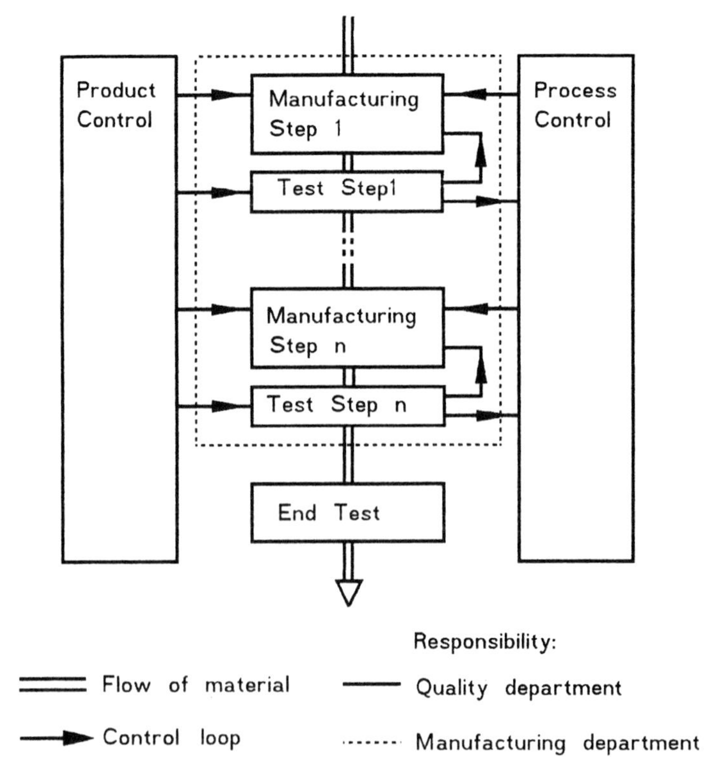

Figure 6.2 Process control and product control – two necessary attributive controlling mechanisms.

Product control comprises:

- control of product quality by attributive tests on a sample basis;
- immediate perception of emerging problems;
- analysis of their causes; and
- counteractive measures and improvement of the processes.

Process control, on the other hand, comprises:

- statistical control of production processes;
- prospective surveillance of all processes; and
- stabilization of these processes.

The goal of quality management is to shift the efforts of quality engineers more and more from product control to process control.

6.2.3 Quality actions

The quality management has to establish definite routines on how to handle requests arising from the above controls. These requests may result from a mismatch between the planned quality goals and the actual measured quality or from a proposal for process improvements. A general overview is given by Fig. 6.3. Severe cases where quality is completely insufficient are considered critical situations and production has to be discontinued. The quality team has to convene an emergency meeting and upper management has to be informed at once. In less critical situations, there is enough time for a corrective loop to come into action, if such a corrective loop has been prepared beforehand. That means measures on how to react to quality problems have to be provided in written form in a failure catalogue and corrective actions automatically executed by the quality engineers. If that is not the case then a quality action has to be initiated. This also happens when a proposal for process improvement is made, if this requires major changes in the product flow.

A quality action has to have a well-defined procedural sequence:

1. an announcement and registration which can be done by anyone who detects a quality problem;

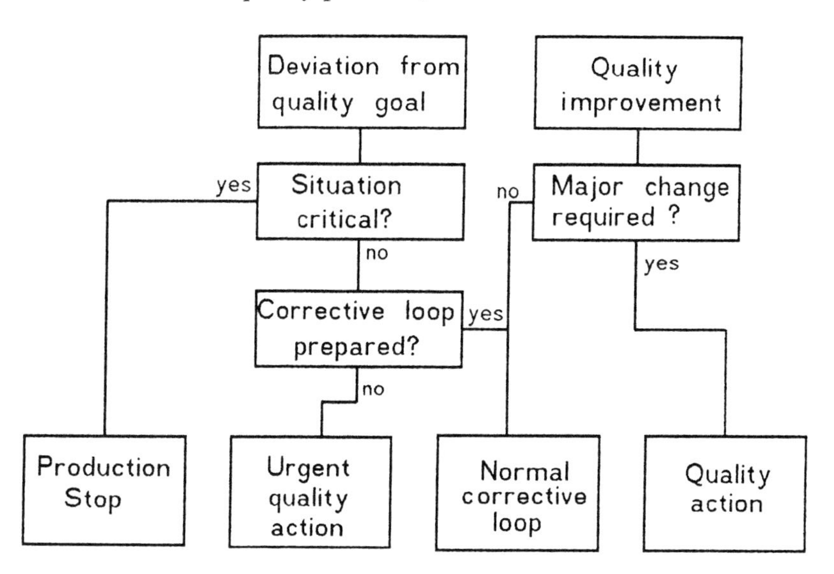

Figure 6.3 Quality action is a defined procedure to solve quality problems and to improve quality.

2. the definition and containment of the problem – the danger is that the problem may expand and the action will never come to an end;
3. the severity of the problem has to be defined;
4. measures have to be taken to cure the deviation of the quality from its goal;
5. the cost of these measures and the time for their execution have to be compared with their profit;
6. the action has to be prioritized and put into an action list; and
7. if the problem is solved then the action has to be closed formally.

An important task for the quality team is to establish a controlling procedure for quality actions:

- Who is responsible for each action?
- Who controls the complete execution of the actions after a predetermined time limit?
- Who controls the failure preventive effect of each action?

6.2.4 Test strategy

Establishing a test strategy is one of the most important facets of quality management. The available knowledge of failure rates, failure modes and their causes, collected in the past, is used now to install an integrated overall test plan. This plan has to view the entire production and test process as one entity. All *ad hoc* designed tests at each production station, developed for a particular process technology and for specific products are replaced by flexible test procedures within an integrated test plan. This does not mean that all tests have to be centralized. The objective is to harmonize the tests to guarantee complete test coverage at least cost, and to avoid double testing. The test strategy has to be structured and hierarchical. General principles are established which are valid from incoming inspection to service. The detailed test planning within each manufacturing step is better performed locally but according to the principles and requirements set up and documented in the test strategy. Under its guidance, the management of the individual test steps shown in Fig. 6.4 and also the test management within a manufacturing line (not shown in Fig. 6.4) will establish their own well-documented test strategies. This is explained in the sections which follow.

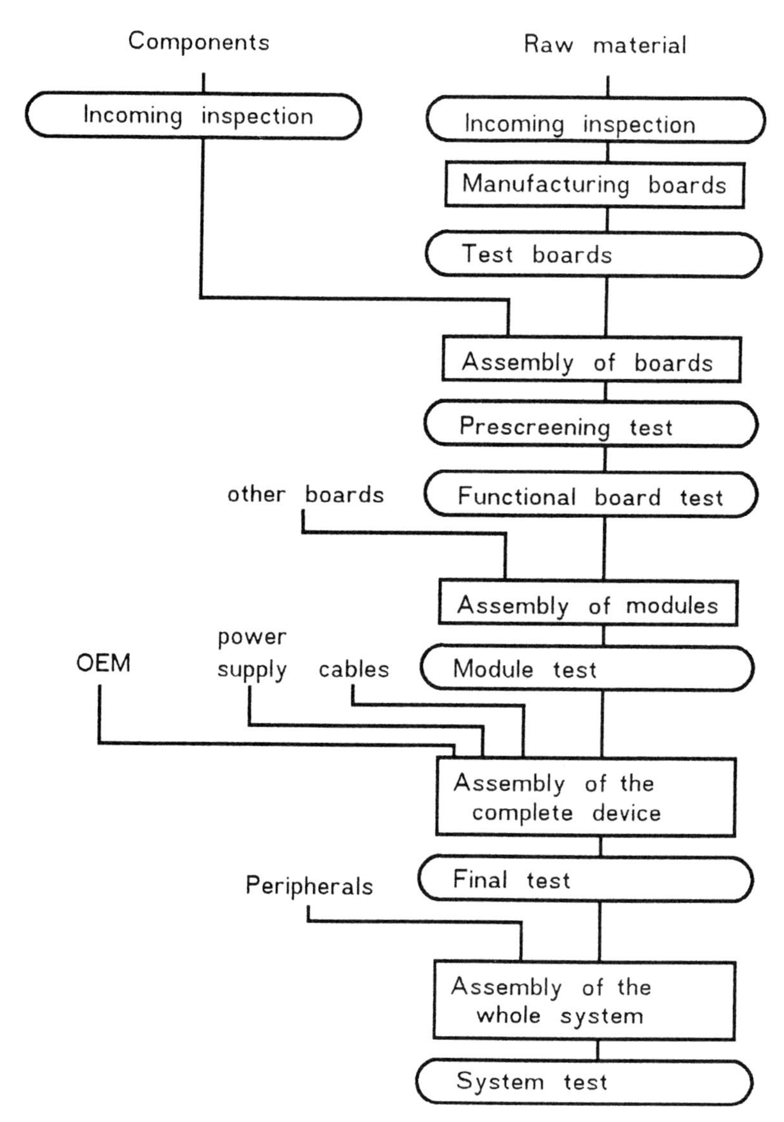

Figure 6.4 Flow of material and test sequence. Immediate tests after each production step are the first measure to avoid failures.

The strategic goals are as follows:

- Selection of the best test plan to get high test coverage at minimal costs. The concept of testing after each production step is established which allows the location of each failure at its origin and avoids double testing at successive test steps. Optimal harmonized test procedures from incoming inspection to final test have to be installed. It is necessary to plan the test coverage to predict failure rates for all test steps. This may include some trade-off between component, board and final tests. An example is given in section 6.7.2.

- Analysis of this preliminary prediction of failure rates and expected product quality. The main failure modes have to be discussed with the design group, then joint actions for design improvements to enhance testability are started.
- Co-operation with the design group to implement design rules for DFT designs (design for testability). A compromise has to be found to get testable products without too much overhead. The quality group has to contribute its testing experience during these negotiations.
- Provision of test tools for automatic test generation with migration capability to subsequent test steps (e.g. from board test to final test and service).
- Installation of fast reacting and effective corrective loops.
- Having determined the test methods and the required test facilities (tester and personnel), obtaining optimal test economy is the next task. This means not only selecting testers, defining the required test time, test personnel and test rooms and the required qualifications of the test personnel, but also calculating the cost of failure prevention, testing and repair and setting up a cost account. The costs for constant training of the test personnel should not be forgotten.

It is important that this strategy is changed if the market conditions or the technological development deem it necessary. Because so many factors influence its structure, computer assistance using artificial intelligence might be helpful in setting up this strategy. But this seems to be a task for the next decade.

How all these goals can be realized for each test step is shown in detail in the next sections, 6.3–6.8.

6.3 TRACEABILITY

A computer aided tracking system identifies samples of each lot of components on arrival, and afterwards tracks them through manufacturing to associate the quality data and the process parameters with the components. Components and assembled printed boards, as well as whole electronic devices and systems are tracked. As shown in Fig. 6.5, which explains the principles of the tracking system, inputs into this system have to come from all production stages. This is made possible by incorporating the manufacturing control into this system.

The tracking system is similar to the ESIS of section 3.2 in so far as it allows a survey of the quality status at any time. Seen in detail, however, both systems are quite different and it is not advisable to mix them. A controlled data link between both systems is a better solution. This tracking system is also different from the documentation of section 6.2.1 because it deals with identifiable physical items and not with corrective actions.

It is important that it is an integrated system which extends over the complete manufacturing process. It has to replace all the separate lists and local databases used by each of the manufacturing and testing locations.

The main contents of this tracking system are:

- all incoming samples or lots of components, marked by an unequivocal identification;
- all printed boards related to the assembled components;
- all electronic devices or systems delivered to the customer with references to the used boards or other components;
- all manufacturing processes with their parameters;
- all test processes with test criteria;
- test results from all test steps including service; and
- results of failure diagnosis and analysis.

The benefits of this system are:

- the failure rate can be made transparent throughout all test steps;
- the effectiveness of correcting measures can be controlled;
- the overall failure costs can be derived specific to each item (component type and vendor, manufacturing process, device or product);
- the results of several test steps can decide the further usability of a sample/lot or board;
- samples/lots which have been rejected several times within a corrective loop can be identified;
- manufacturing-related failures can be traced to earlier stages of production;
- failure clusters which become apparent in several later stages can be recognized;

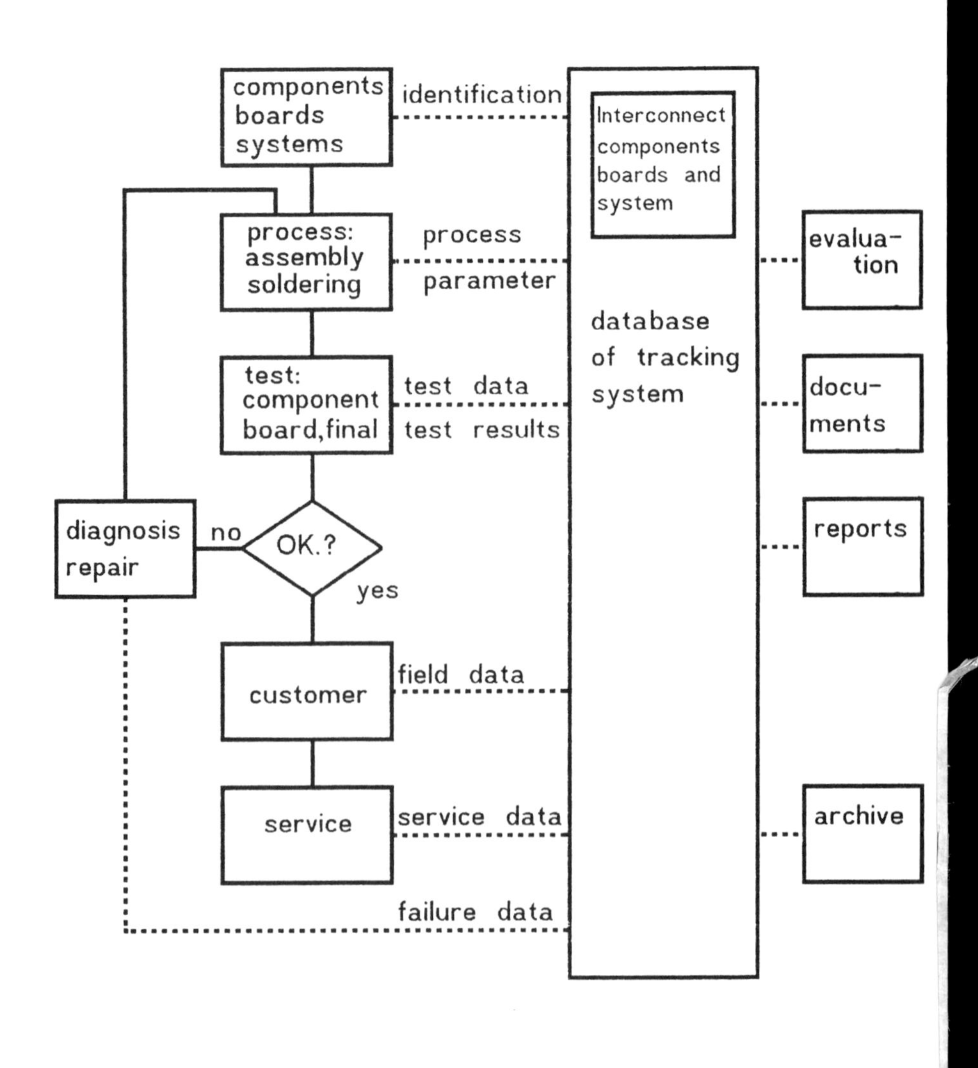

Figure 6.5 Traceability – an integrated tracking system for all items during the complete manufacturing process is the basis for an effective quality management.

- customer failures can be traced back anywhere from service to manufacturing;
- documentation of all quality data specific to each entity (quality record) becomes possible; and
- in the case of product liability actions, the mandatory retracing of all components to the concerned manufacturing places is possible.

To be adequate, the tracking system has to be able to:

- store huge quantities of data;
- permit simple and effective data input (bar code);
- permit flexible and comfortable data retrieval;
- be adaptive to future request; and
- feed a long-time archive (about 10 years) with a fast and secure data retrieval.

6.4 INCOMING INSPECTION

6.4.1 Management of incoming inspection tests

The purpose of incoming inspection is to ensure a low failure rate for all components which are fed into the manufacturing line. Initially, a 100% test of all incoming components has to be performed in order to obtain an overview and knowledge of the actual quality status of components. This is a necessary step on the way to reducing these tests to a final ship-to-stock procedure. The conventional sample test to guarantee a certain AQL value is not sufficient; a failure rate in the region of dpm is required. The failure rate can be reduced thus far only by 100% testing, failure analysis and close co-operation with the vendor. This is shown later in Fig. 6.26, which has to be seen in conjunction with Fig. 6.27 showing the advent and the expiry of 100% testing, the constantly rising number of components bought under a ship-to-stock contract, and the simultaneously decreasing dpm value. Even when a 100% test can be omitted, the quality has to be monitored by sample tests. In the last period shown in Fig. 6.27, the sample tests increased slowly as a consequence of this monitoring. This is proof that one of the goals leading towards good quality at low cost can be reached. It should be noted that Fig. 6.27 refers only to ICs. Activities for a 100% test of passive components have not been undertaken. Proper management of incoming inspection is the key to keeping the cost of incoming inspection as low as possible and being prepared for future requests. Complete traceability of all parts during the production is vital, as explained in the preceding section. To estimate the necessary test capability, to buy the right tester and, last but not least, to initiate the failure analysis and correlate the result with the vendor, are all the responsibility of incoming inspection management.

This management should establish an immediate flow of materials inside the area of incoming inspection, from the delivery of components at the front door of the factory to the test facilities, and then from there to an intermediate store, and from this store to the production line. Any crowding of material in this flow should be avoided. This will be facilitated by an appropriate spatial grouping of all working places concerned, from unpacking and checking of the contents of delivery up to the testing area. A strict first-in, first-out principle within the store ensures that no component sinks to the bottom.

The tracking system described automatically gives instructions to the test engineer about the amount and kind of testing for each lot. This information is entered into the system by the quality engineers and so made available to the test engineers. The test engineers in turn store the measurement results in the system. Because of the huge amount of numerical data, graphical presentation is important. Any change in quality and pointers to weak types or vendors must show up. The tracking system also supervises the content of the component store and shows any irregularity in delivery.

As soon as quality has improved, this system allows the management to shift to a ship-to-stock procedure just by reducing or omitting tests on a typewise or lotwise basis. This can be done by a mere change of the test instructions. Failure rates later in production have to be traced back to lot number and date code in order to start failure preventive actions, for instance a re-establishment of 100% testing. Further, a shift to a ship-to-line procedure is possible through a gradual reduction of the admissable content of the store by a software switch. It is important that quality management is flexible and can respond to changing requests. The tracking system will be a great help in this.

6.4.2 How to choose the right testing hardware

Classification of testing hardware

There are many different choices when selecting testing hardware for incoming inspection, from cheap go–no go testers to dedicated automatic testers. Testers can be divided into three groups according to price. Table 6.1 shows the cost of investment and operation of these testers.

Table 6.1 Cost of test equipment

	US$
Investment	
automatic tester	0.2 – 1 million
benchtop tester	5 – 100 thousand
go–no go	2 – 15 thousand
Operating cost	10% of investment per year
Test personnel	
Depreciation	
Cost of test programs including DUT board	
SSI/MSI	1 – 2 thousand
LSI	2 – 10 thousand
VLSI	10 – 20 thousand
ULSI	20–150 thousand
Cost of failure analysis	
Cost of correlation and statistics	

1. **Go–no go testers** These are cheap and simple functional testers which will show whether a part will work. Measurements can be made of some DC parameters at an elevated supply voltage or temperature. High-speed functional tests can be made to a limited degree. Testers which compare the device under test to known good samples also belong to this group. Go–no go testers are most suited to manufacturing or incoming inspection tests of less complex components. They are less suitable for failure analysis and evaluation of complex circuits.
2. **Benchtop testers** This is test equipment consisting of several off-the-shelf test instruments which are combined to perform the desired tests. For instance, several pulse generators, an oscilloscope, a plotter or printer, a programmable power source and a personal computer can be connected together to test dynamic parameters of rather complex circuits. Usually an IEC488 bus is used to connect all parts. These combinations are very flexible because all kinds of instrument can be switched together and each instrument can be exchanged separately. The drawback of this solution is the necessity of software and higher qualified operating personnel. These test installations are usually

composed of single standard off-the-shelf instruments with software written by the user, but there are some tester manufacturers who offer complete solutions including software. Some specialized software companies also offer complete and very flexible software solutions for a choice of recommended hardware, including automated range settings and reconfiguration of the instruments using the personal computer and including data presentation.

3. **Automatic testers (ATE)** These are dedicated testers sold as a way of testing all kinds of digital components up to very complex microprocessors and gate arrays. ATEs can have either a shared resource or a resource per pin architecture. They differ in the way resources such as timing generators and formatters are connected to the pin electronics (Fig. 6.6). Shared resource testers multiplex a limited number of precise and costly resources to electronics with a larger number of pins. Because the path delay from master clock to the pins depends on the routeing through the switching matrix, it is more difficult to compensate the resulting deviation by software. These problems increase with pin count.

   For resource per pin testers, the resources are dedicated to each tester channel and no switching matrix is necessary. The timing adjustment is easier and the timing accuracy of the system should be higher, but the cost per channel is increased. Users of both kinds of testers came to the conclusion that overall the pros and cons balance. Nevertheless, the resource per pin testers seem to be increasing their market share slowly. That comes from the better integration of the hardware, which reduces the cost gap, and from advantages in test program development (Barber and Satre, 1987; Stover, 1984).

   Dedicated ATE test systems for RAMs for analogue circuits as well as for telecommunication circuits are offered too. Complex circuits, gate arrays which contain large parts of digital and analogue logic on the same chip are difficult to test. Strict separation of both kinds of logic is recommended by added enable inputs and test outputs. Nevertheless, this is an open problem, because the number of such circuits may rise in future.

   Asic design verification systems (DVS) are a special kind of tester, lying between go–no go testers and ATE. They are cheaper than ATEs – about a tenth of the price. High pin count is not a problem for these testers and they can handle test patterns which come from simulators. Their strength lies in detecting

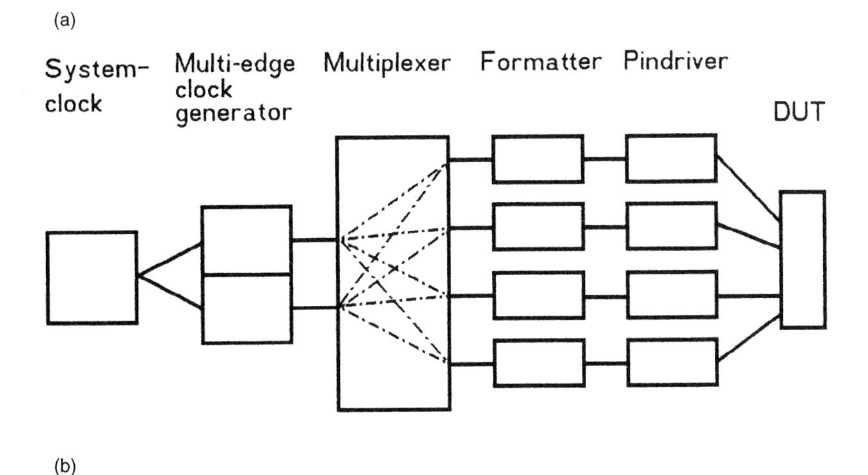

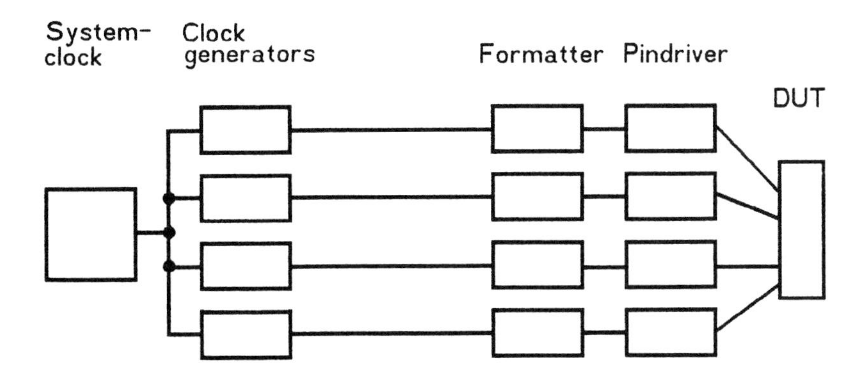

Figure 6.6 Two different tester architectures: (a) shared resource; and (b) resource per pin.

logic bugs in asic designs, so they are useful to the designer of asics, in the development phase. But the flexibility and accuracy of their timing sources and their DC measurement capabilities are very limited. Therefore they are less suited to evaluation tests or to test circuits with a more complex timing requirement like microprocessors. Manufacturers of these systems aim to improve and enhance their testers and so overcome their limitations. They may yet become important competitors to ATE testers for applications where asics are the main test object.

Required test capacity

To select adequate testing hardware for your company, the following questions have to be answered:

- How much test capacity do you really need to ensure the quality of your products?
- What kind of testing hardware will best fit your requirements?

The primary reason for a user of electronic components buying test equipment is to perform incoming inspection. For components which you buy from a qualified vendor under a ship-to-stock contract, the amount of incoming inspection can be decreased considerably without loss of quality, as shown in Chapter 7. However, until all your vendors are qualified, you still need 100% incoming inspection according to the rule that it is cheapest to detect failures at the earliest stage of production. This also holds true for new component types which might contain technological risks. At the moment, asics seem to be one of those components, because they have a tremendous increase in integration density and in technological progress whereas the production volume is rather small. The fault coverage of the vendor's test programs is not yet rigorous enough to detect all failures.

In addition, even the quality of components bought from qualified vendors has to be monitored by sample tests. So, although the amount of incoming tests should be decreased as much as possible, it is not advisable to relinquish completely the ability to perform such tests. This would allow the user to become helpless in the case of a sudden quality breakdown.

Test equipment is also necessary for users of more complex components if they have decided to do failure analysis. For diagnosing and analysing failing microprocessors or asics, automatic testers are essential. This allows you to talk with competence to your vendor and so help to improve the quality of these circuits.

Last but not least, automatic testers are mandatory or cost-effective in the evaluation of complex circuits as explained earlier (section 4.4).

So, the choice of testers is driven by two demands: the tester has to be technologically adequate for the test objects and economically adequate for the test volume. Because of the first demand, automatic testers are recommended to manufacturers of electronic systems that use ICs of high complexity or advanced technology. For companies using less advanced but still complex components like microprocessors or telecommunication circuits (e.g. manufacturers of PCs), it is a question of balancing the investment in an automatic tester against the cost of skilled operators necessary when using benchtop testers.

Even for smaller companies with a broad spectrum of products, it is usually more economical to use at least one automatic tester. It can be shared by the evaluation group during the daytime and by the factory for incoming inspection overnight. It may prove economical to use additional benchtop testers or go–no go testers for special test problems and for simple components, but bear in mind that incoming inspection is changing. Less volume, more intelligence and a high degree of automation will be required in future.

How to select the right vendor

If you have made a decision to invest in an automatic tester or some other comprehensive test equipment then select the right vendor with a tester that best fits your needs. For this, you should take the following five points into consideration:

1. The vendor should be among the leading manufacturers of testers with a considerable market share. This is to ensure support, maintenance and a potential upgrade during the lifetime of the product. A vendor offering a variety of different testers (evaluation and production testers) is an indication of professionalism. A smaller vendor with aggressive marketing may offer you more advantageous conditions and possibly the most advanced technology but you must always be aware of the risks of buying from a newcomer.
2. The tester you buy is always made using components which are one step behind the latest technology on the market. This is because it takes some time for the vendor to qualify, design and produce a tester. Take care that the tester is not made using totally obsolete components or else the tester may soon become obsolete itself.
3. The vendor of a tester should be willing and capable of providing complete documentation. A small user guide is absolutely necessary but not enough. It should be complemented by detailed documentation for operation, calibration and basic maintenance. The user should be able to replace parts accessible to the operating personnel, like pin electronics.
4. The vendor should offer a chain of training courses (basic and advanced) and also, if necessary, maintenance courses.

5. The vendor should offer a choice of different maintenance services:

   (a) A complete maintenance with regular calibration and a written guarantee for a certain minimum down time. This may cost more than 10% of the purchase price per year.
   (b) A partial maintenance if the user is willing to do calibration and some basic maintenance himself. But the contract should guarantee that in the case of severe trouble, the vendor can be called in. The drawback is a potentially longer response time in these cases; this may prove a good compromise for lower costs.
   (c) The user trains his own people to do the maintenance. In this case, the vendor should be liable for the provision of spare parts during the lifetime of his products. This method is only viable if the user has several testers of the same kind in operation.

Evaluation of test equipment

After selecting one or more potential vendors for your test equipment, the next step is a thorough comparison of their testers. You must find out which tester best fits your needs, not only for the present but also for the future. Bear evaluation tests in mind as well. Start with an examination of the datasheets. In Appendix E, a questionnaire shows which items have to be considered when you are selecting a rather expensive ATE system. For less expensive test equipment, this questionnaire can be reduced accordingly. Next, a specimen of the tester has to be examined technologically either at the customer's or at the vendor's site. Oscillographic pictures have to be taken from all output signals at critical conditions. Figure 6.7 is a sketch of a multi-exposure photograph taken from the output of a digital pulse generator. The pulse delay was increased by one step between each succeeding exposure. A timing non-linearity is observed when the range is changed. The inputs have to be checked too.

When examining ATE systems, typical test objects of different kinds should be given to the vendor to carry out delay measurements. The samples have to be tested manually before and after these measurements. Comparing the results gives an impression of the tester's accuracy.

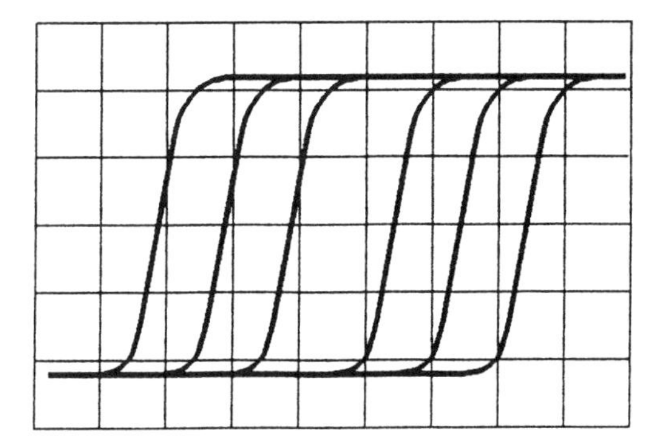

Figure 6.7 Evaluation of test equipment: sketch of a multi-exposure oscilloscope photograph showing a nonlinearity of time setting at range change.

Checking the timing accuracy of a tester is a special problem. Resolution and accuracy data given in datasheets are often unclear or misleading. Resolution means the smallest increment or decrement possible while setting a timing parameter. It is easily specified and checked. Of more interest is the accuracy, i.e. the difference between the value set or read on the operator panel and the real value of an output or input. The timing accuracy is especially difficult to check for the customer because the definition in the datasheet is often vague. The vendor either specifies the accuracy of its timing system separately for each of its parts or gives an overall accuracy without specifying how it is defined and measured.

The standard method for checking the accuracy is to measure external timing standards, like calibrated delay lines. The problem with this method lies in the fact that the standards are not available everywhere. Delay lines have to be terminated and are used in well-defined conditions, whereas the real test objects are of high impedance and are mounted in a less ideal manner. A proven method is to use the inherent measurement capability of an ATE system itself for checking. This gives you a feeling for the consistency of the measured values. So you are independent from outside standards or instruments and you can perform this test on a demonstration tester at the vendor's site too. All drivers are connected in turn to all receivers manually using one and the same coaxial cable of a definite length. A sequence of delay measurements is made for each driver–comparator pair using different timing generators, time settings and formats. From the results, the following items which contribute to the overall accuracy can be derived:

1. the linearity of timing generators
2. skew between different timing generators for drivers
3. skew between different timing generators for comparators
4. format to format skew
5. reproducibility
6. skew between drivers
7. skew between comparators
8. skew between comparator levels (low and high).

The overall accuracy O_{AA} can then be estimated using the above listed skews F_n by the failure propagation law:

$$O_{AA} = 2 \times \sqrt{\left(\Sigma\left[\frac{F_n}{2}\right]^2\right)}$$

All this, except the attachment of the cable, is performed automatically by a check program running on the tester. An example of the results of such an evaluation, which was performed on comparable shared resource testers of three different vendors with 256 pins each, is given in Table 6.2.

Table 6.2 Example of tester accuracies obtained by evaluation

Tester	*Accuracy*		*Safety margin*
	Specified (ps)	*Measured (ps)*	*(%)*
Vendor 1	1200	560	53
Vendor 2	1200	1160	3
Vendor 3	1400	800	42

This result shows that the specified value of Vendor 2 is a typical rather than a maximum one, whereas Vendors 1 and 3 have some safety margin to the specified values.

The DUT board or loadboard is an essential part of a tester. It contributes noticeably to the total cost of testing because a new DUT board is regularly required to test each new component, not forgetting a spare board as a back-up. So when you are evaluating testers, check the performance of the DUT board too.

- Is it easy to mount on the test head?
- Is it cheap and uncomplicated to manufacture?
- Is its electrical performance adequate for the test objects?

The pin count of advanced packages may exceed the number of channels of a tester. As long as this situation only occurs on a few circuits, a two- or three-pass test is the cheapest solution. Whereas all input and bi-directional pins of the test object have to be connected to a tester channel for each pass, the output pins can be connected partially in turn. So, more pins can be tested than there are channels available.

$$p = c + s + \frac{o}{n}$$

An example explains this:

Tester channels	c	256
Supply pins	s	28
Logic pins	l	292
Input pins	i	219
Output pins	o	73 (37 + 36 for $n = 2$, two passes)
Package pins	p	320

6.4.3 How to procure test software

To write a test program for incoming inspection of SSI/MSI components is very easy and cheap for the customer. It becomes still easier by using a test program generator. Such program generators are offered by most tester manufacturers or by software companies. It runs on a workstation or on the tester itself. The necessary input from the user consists of static and dynamic datasheet parameters; the output is a complete test program which in most cases is error free at the first run. About 1–2 days are needed to create and debug such a program.

To write test programs for complex LSI circuits like microprocessors is more expensive and time-consuming. It may need 1–6 months for a trained customer to write and debug such a program alone. There are three other methods of procuring a test program for LSI components:

1. Obtain them from the vendor of the tester. Some vendors will provide test programs for some commonly used circuits, like standard microprocessors and peripherals.
2. Obtain them from the vendors of the circuits, because they need a test program for their own outgoing inspection. Many vendors, however, hesitate to give their test program to the customer. Furthermore, it is difficult to know to what extent the logic function is tested by the program. The vendor knows the particular weak points, so a short test time might be his primary request.
3. Buy a test program from specialized software houses. These companies sell their programs to several customers, therefore they can be cheaper than writing a program yourself. In addition, you can get guarantees and service from them. But be sure that the installation of the program at your tester by this company is included in your contract. For any program upgrade which may become necessary on revision of the circuit, you have to pay an extra fee.

Test programs for asics pose special problems. The vendor of asics demands from his customer a simulated test pattern together with the wiring list. After it has passed a resimulation at the vendor's site, this test pattern becomes part of the development contract. All parts which pass this test pattern, will be good parts in the view of the vendor, even if they do not run in the customer's application. Therefore it is up to the customer to make sure that the fault coverage of the pattern is high enough; about 95–98% for static stuck-at failures. It is good practice to insert into the contract the right to expand the pattern later on at no cost, as long as they do not contradict the wiring list. Sometimes, vendors of asics offer to derive the test pattern from the wiring list by simulation with a fault simulator like LASAR. So problems which may arise from differences between the vendor's and the customer's simulators are eliminated and the customer may save the cost of fault simulation. The vendor promises to deliver fully tested components to the customer. Again the customer has to make sure that the fault coverage is sufficient to avoid problems in production.

Further consideration of the principles of test program generation is given in Appendix F.

6.4.4 To what extent do components have to be tested?

There is a certain danger of the quality manager falling between two stools when defining the scope and the extent of incoming inspection. The production manager and the maintenance people on one hand put pressure on him to test many parameters at tight limits. This should improve the reliability of the end product and reduce maintenance costs. Sometimes, even tighter limits are demanded than granted by the vendor's datasheet.

This transition from datasheet control to selection test is not the right way to improve quality because it leads away from the goal of ship-to-stock and from cost reduction. On the contrary, all components should be used well below their datasheet limits. In co-operation with the vendor, a graded scale of test limits is established which reduces rejections to a minimum and provide a safeguard in applications.

Vendor's outgoing test: tight limits
Datasheet limits: less tight limits
Incoming inspection: less tight limits
Customer's application: least tight limits

Any exception to this philosophy must be urgently required by circumstances.

An example is the maximum operating frequency of a component like a microprocessor. Suppose the vendor specifies this frequency in his datasheet at 50 MHz. Then the customer's design manager will be forced by competition to use the component at exactly this frequency, without caring about the test conditions in the vendor's datasheet. These may be nominal conditions which do not reflect at all the conditions of real usage. So the design manager may demand an incoming inspection at 55 MHz to guarantee the 50 MHz operating frequency. But this means increased cost for testing and for rejected parts. It is up to the quality manager to steer between these two rocks. He or she has to co-operate with the vendor to choose incoming inspection test conditions that make best use of the available margin between the outgoing test conditions at the vendor's site and the conditions of the real usage of the component.

Some manufacturers of electronic devices or systems submit their outgoing products to a marginal check at end test. This means that the products are tested in conditions which are more stringent than normal operation, which may be larger supply voltage tolerances, higher frequency or temperature. The purpose is to screen out samples which may fail later during the warranty period. Because failure detection is cheaper the earlier it is performed, the quality manager is asked to tighten the limits of incoming inspection as well. Again, a way out of this problem could lie in co-operation between vendor and customer. Both should agree to solve this problem stepwise. As a first step, the customer can perform tighter incoming inspection and the vendor can agree to accept a certain percentage of rejects without retest. So the risk is split between both parties and calculable for the vendor. As a second step, the number of rejects is noted and the vendor has a chance to improve the test margin of the components. The goal is to come to a negligible number of rejects. Then a ship-to-stock agreement will be negotiated.

Testing dynamic parameters of asics has its own problems. Most vendors of asics specify only typical values for cell delays and some kind of tolerance factor max/typ, both being derived from simulation. For a very limited number of paths per asic, a maximum delay value can be specified in the development contract. It is advisable to design in and to integrate a special test path used only for dynamic tests. It has to provide a long chain of gates which permits well-defined delay measurements. Both vendor and customer should agree on delay limits for this path. Some selectability can also be provided by this measure.

As far as memories are concerned, the increased quality means this item is a good candidate for a ship-to-stock contract. Incoming inspection is only necessary for new advanced circuits. The ever-increasing bit count makes it more and more difficult to perform a complete test (for instance with galpats) as has been done with smaller memories in the past. Today, a marching pattern or butterfly pattern have to be sufficient. In any case, the vendor has to support you with the assignment of row and columns to the address pins.

It was stated at the beginning of this chapter that 100% incoming inspection is an intermediate, but necessary, step on the way to reducing costs. It has to be replaced by a ship-to-stock procedure supported by quality monitoring as soon as the incoming quality becomes sufficient. However, all the aforementioned problems in this subsection have to be solved. Without this, the cost reduction which is aspired to cannot be achieved.

6.5 BOARD TEST STRATEGY

6.5.1 Procedure for choosing the optimal test strategy

As soon as the failure rate of the components diminishes, incoming inspection loses its importance and board testing becomes the central aspect of quality assurance. The goal of board test management is to choose the right test strategy to achieve the desired board quality from boards with a known failure rate, at a minimum of cost. This is explained in Fig 6.8.

At first, the failure rate of the test objects, the boards to be tested, has to be derived. The second step is to find a suitable test strategy, i.e. either one or a hierarchical sequence of test proceedings with different testers and test methods. This also depends on the types of components which are assembled on the board; see example in Table 6.3.

The next step is to calculate the quality which will be achieved by the chosen test proceedings and to calculate the cost for this

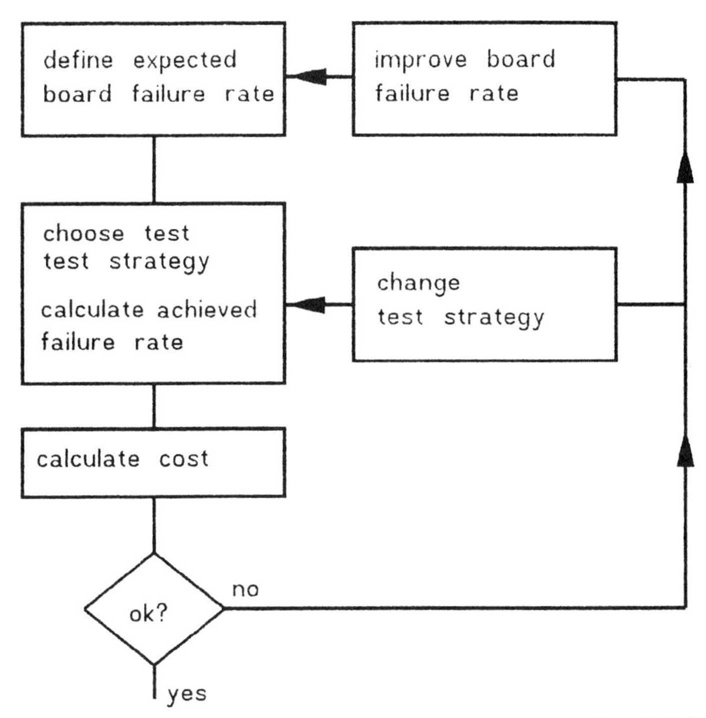

Figure 6.8 Board test management: a procedure for achieving high board quality with an economic board test.

(Table 6.5). If the result is not acceptable then a change in the test strategy should be tried. Other versions or combinations of test proceedings should be chosen. If this is not successful then the failure rate and/or the testability of the boards has to be improved, as sketched in Fig. 6.8.

It is very important that this procedure is repeated every time the preconditions are changed, i.e. when the failure rate of the components has decreased or the quality of the manufacturing process has improved. If this is the case then the cost/performance optimum of the test proceedings will also have changed and cheaper tests may be sufficient. The same is true for the opposite case. If the final test showed an increasing rate of board failures, then the test coverage of the board tests has to be improved. Therefore, it is recommended that the procedure is software assisted so as to be repeated easily. A computer program performs all the calculations which are described in the following sections.

Estimation of failure rate

The failure rate of a printed board after assembly is derived from the failure rate of all components and from the failure rate of the manufacturing process steps. The component failure rate comes from the preceding incoming inspection or from the vendor's data. Modern boards are assembled with asics, microprocessors and peripherals, RAMs, a small quantity of MSI circuits to drive an interconnection bus or cables and passive components like blocking capacitors or termination resistors. In the past, boards consisted of many MSI/SSI circuits and analogue circuits as well. If all these components have a low failure rate due to 100% incoming inspection or to an equivalent ship-to-stock procedure and if the boards are pretested to omit open or shorted wires, then most of the failures detected by board or module tests are manufacturing failures. These are mostly bad soldering and tin bridges (more than 80%). This is demonstrated in the following example.

A completely assembled printed board contains:

10 LSI circuits (asics)	with	250 pins	and	1000 dpm:	10 000dpm
32 RAM circuits	with	16 pins	and	100 dpm:	3200 dpm
20 MSI circuits	with	16 pins	and	50 dpm:	1000 dpm
20 passive components	with	2 pins	and	20 dpm:	400 dpm
82 parts	with	3372 pins			14 600 dpm

Taking into account 3372 solder points and about 10% additional possible production-related faults and a failure rate of 20 dpm results in 3700 solder points with 20 dpm: 74 000 dpm. The total failure rate of the board is 88 600 dpm. That means that 8.86% of all boards on average are faulty and 83.5% of all failures found during board tests in this example are caused by failures during manufacturing. If microprocessors are used instead of asics, this percentage will be even higher. It should be noted that the failure rates in this example are somewhat prospective. At the moment, the failure rate of RAMs can rise to 500 dpm and of solder points to 50–70 dpm.

Board test methods and board testers

The following are descriptions of some important test methods for testing printed boards (Bardell *et al.*, 1987; Eichelberger and Williams, 1973; Flaherty, 1993c; Frank and Sproull, 1981; Gruetzner and Starke, 1993; Johnson, 1993; Markowitz, 1992; Mizko, 1986; Parker, 1992; Ratford, 1992; Romanchik, 1993c; Savage *et al.*, 1993; Scheiber, 1992; Steward, 1977; Totton, 1985).

1. **Visual or optical test**
   Today, as in the past, visual test means an inspection of printed circuit boards using human eyes with or without optical magnification. Wrong or wrongly mounted components and bad soldering can be detected. Optical testing in future might mean an AOI (automatic optical inspection) system, using three-dimensional optical image or laser scan to inspect solder points automatically. Transmission radiography using X-rays is another inspection method of the future. Neither an adaptor nor a test program is required, so the test is relatively cheap. Some restrictions are to be observed in the mounting density of components. At the moment, however, the image processing software allows automatic failure detection on very regular assembled boards only, otherwise the interpretation has to be done manually.
2. **Prescreening**
   Besides an optical test, a simple analogue in-circuit test can be used to perform prescreening to detect and eliminate bad soldering. Missing or wrong resistors can be detected too. A needle adaptor is needed for this test. Because this test is relatively cheap, it is often used to save test time on high-performance testers. It is wise to start the test using low voltages so that potential shorts will do no harm.

3. **In-circuit tester**
   Digital and analogue components are tested by a needle adaptor which contacts their input and output pins. The dangerous backdriving problem (forcing a high potential to a low output) is solved by shortening the test time. Static testing and to a limited degree dynamic testing too is possible. Deriving test programs can be facilitated by learning from a known good board. A challenge and a limitation for the use of ICT is the pin density of modern boards. It increased from 6 pins/cm^2 with 600 µm line width to 15 pins/cm^2 with 180 µm line width for double-sided boards with SMD components. This leads to adaptors with several thousand needles. Both sides of a double-sided board usually have to be contacted at the same time.
4. **Digital function test**
   Test patterns are applied and the response is measured on the board connectors. To set up the test program by simulation of the complete board is often very time consuming and requires simulation models for all components. It is sometimes difficult to obtain models from the vendors for new components. Due to the increasing logic content of boards, a sufficient test coverage is hardly achievable because of the pin limitation of the board connector. The application of this test method on telecommunication boards may become problematic. Fault location is often still performed manually by guided probes. Despite these drawbacks, the function test is required to perform dynamic tests and has become more or less a standard for high-performance board tests.
5. **Substitution or emulation test**
   This test is limited and especially suited to microprocessor boards. The processor or the RAM is replaced by the same function inside the tester. The processor on the board under test has to be socketed or a clip adaptor used.
6. **Self test**
   This is an alternative test method for testing processor boards, e.g. of PCs. A test program is loaded into the RAM on the board of the test object. No special hardware is required but fault location is difficult.
7. **Level-sensitive scan design (LSSD)**
   This is special hardware designed into asics to ease testing at incoming inspection. It can be used for board testing too, especially in combination with a boundary scan. LSSD requires additional hardware built into all sequential elements by which

they can be connected in series to form a shift register called the scan path. Data can be shifted in to load all flipflops and the contents of all flipflops can be shifted out for analysis. So, it is very simple and straightforward to create a test program automatically but the number of test patterns is rather high. The LSSD method is not commonly standardized and it seems that each manufacturer has his own. Depending on the kind of LSSD, the gate count of each flipflop is increased by 20–50% and the use of simple latches has to be avoided. The drawback is that four additional pins are required for each asic together with a hardware overhead of 10–20% for the complete electronic system.

8. **Boundary scan**

This new test method, specified by IEEE 1149.1, can be used to test circuits, boards and systems as well. By boundary scan, the correct function of components (internal test mode) and correct chip-to-chip wiring (external test mode) can be tested. Using the external test mode, up to 95% of all manufacturing faults can be detected. Besides these modes, a sample mode exists, by which the signals on all component pins at a certain point in time can be shifted out for analysis.

Boundary scan requires components which are specially prepared for its use by design. This is not a problem as far as asics are concerned if a hardware overhead of 5–15% (about 130 gates plus 10–15 gates/pin) and four additional package pins are considered. For PALs, an overhead of up to 50% has to be taken into account. Standard circuits with integrated boundary scan are still rare today but their number is increasing rapidly. The latest microprocessors include this feature and RAMs with boundary scan have been announced. Bus drivers with integrated boundary scan are available from a few vendors.

Boundary scan does have some drawbacks. Because it is purely static, no dynamic problems can be solved. In addition, some performance penalty has to be paid (about two gate delays), and it is difficult to detect problems on power and ground pins.

Nevertheless, the acceptance of boundary scan is rising; some new board testers incorporate boundary scan test capabilities. The equipment costs are low, no needle adaptor is required and the programming time is shorter than for function tests. Boundary scan, because it is integrated into the components, can be used for both final testing and maintenance, which increases its cost efficiency. Although boundary scan is standardized, there is not much standardized software for test program generation available.

9. **Built-in self test (BIST)**
   BIST can be realized at chip level or at board level. By additional hardware on-chip (pattern generator and signature analysis), BIST makes the test of components very easy. The same restrictions for boundary scan are also valid for BIST: referring to standard circuits, this feature is not often available today; in future. BIST may become the test for asics on board. With an additional test processor on board, a self test at board level can be performed (see section 6.5.4).
10. **System test**
   The board is placed in a real system and special test software or system software is running. No special hardware is required but fault location is often extremely difficult. This test is used mainly for consumer products such as PCs.

Predicted board quality

A hierarchical sequence of test procedures can be installed to optimize board testing. The following sequence is given as an example of a more conventional test strategy:

1. optical inspection or prescreening to find opens and shorts;
2. test driver components and wiring by ICT; and
3. test all LSI components and RAM by a dynamic functional test.

Table 6.3 The diversification of board failures

Failure mode	*%*	*dpm*
Open/shorts/misplacements	83	72200
Resistors	0.4	350
Capacitance	0.1	50
Static failures	6.4	5600
Dynamic failures	7.1	6200
Temperature-dependent fails	2.5	2200
Voltage-dependent failures	0.5	400
	100	87000

Column 2 of this table shows the percentage of board failures for the different faiure modes listed in column 1. In column 3 these values, which were obtained by experience, are applied to the example of section 4.5.1 with a failure rate of 87 000 dpm.

Many other combinations are possible. AOI, boundary scan and BIST may become the sequence of the future.

To calculate the final board quality which can be achieved by this test strategy, first the component and manufacturing related failures have to be assigned to the failure detecting properties of testers. This has to be done using data obtained from experience (Table 6.3).

The failure modes listed in the first column of Table 6.3 are a simple example. They can be more diverse, for instance: opens at board, opens at connectors, opens at pins, function failure of RAMs etc., in order to give a more exact basis for the calculations which follow.

The next step is to determine a test or a sequence of tests which is appropriate for eliminating the above failures. For each single test, the probability of detecting the failure modes in Table 6.3 has to be derived. In Table 6.4, an overview is given of which failure modes can be detected by which testers. The probability depends strongly on the tools available and requires a lot of experience to arrive at valid numbers. For ICT and function test, it depends on the models available for the circuits and on the simulation tools for deriving the test programs. The test coverage of boundary scan depends on the percentage of parts with boundary scan features.

For each test step, the resulting failure rate, after the test has been performed, is calculated (Table 6.5). In this example, a pretest is performed, then an in-circuit test followed by a function test. In columns 3, 5 and 7 the probabilities of these testers detecting the failure types are listed, and columns 4, 6 and 8 list the resulting failure rates after performing these tests. So the final board quality is derived, which is about two magnitudes better than the initial quality.

It can be seen that simple failures are detected to a high degree whereas dynamic or temperature-dependent failures are more difficult to detect. This is an input when improving the test sequence if the aspired goal is not reached.

Cost estimation

The last step is to calculate the cost of board testing. This consists of costs independent of board quality and of failure-related costs.

The cost per board independent of failure rate is given by:

Test time per board
× (wages + depreciation of test investment)
\+ (cost of adaptor, software and maintenance)
/ number of boards

Table 6.4 Degree of failure detection by different test procedures

	Failure mode							
	Visual	*Pretest*	*ICT*	*Functio n*	*Self test*	*Boundar y*	*BIST*	*System*
Open/shorts/misplacements	X	X	X	X	X	X		X
Resistors	X	X	X	(X)	(X)			(X)
Capacitance+	X			(X)				(X)
Static failures			X	X	X	X	X	X
Dynamic failures			(X)	(X)	(X)		(X)	X
Temperature-dependent failures*				(X)				(X)
Voltage-dependent failures				(X)				(X)

X = most failures detected (>50%).

(X) = some failures detected (<50%).

* Tests at elevated temperature are required to detect temperature-dependent failures.

\+ Defect block capacitors can only be detected by optical inspection, misplaced capacitors to a limited degree by functional tests.

Table 6.5 Decrease of board failure rate by a sequence of test steps

Failure mode		*Step 1 Pretest*		*Step 2 ICT*		*Step 3 Function*	
		%	*dpm*	*%*	*dpm*	*%*	*dpm*
Open/shorts	72200	90	7220	90	722	99	7.2
Resistors	350	10	315	90	32	90	3.2
Capacitance	50	0	50	0	50	10	4.5
Static failures	5600	0	5600	90	560	99	5.6
Dynamic failures	6200	0	6200	20	4960	90	496
Temperature failures	2200	0	2200	0	2200	90	220
Voltage failures	400	0	400	0	400	90	40
Total	87000		21985		8924		776

For ICT and function test, failure models of all circuits were available. Function test at elevated temperature had been added.

The cost per board dependent on failure rate is given by:

Repair time / failure
× (wages + depreciation of tools)
× number of failures/board
+ time for retest per board
× (wages + depreciation of test investment)
× percentage of faulty boards

These have to be added and compared to the expected goals. If the goals are not reached (the cost is too high or quality is too bad) then this procedure has to be repeated using another sequence of tests or with an improvement in failure coverage.

If no satisfactory solution can be found, then the incoming board quality has to be improved further, as indicated by the return branches of Fig. 6.8.

6.5.2 Execution of board tests

Selection of board testers

After a decision has been made on an optimal board test strategy, the test equipment is installed and the tests performed. The selection of test equipment for board tests is as important as for ATE testers. The same principles for the selection of component testers govern board testers too. When planning the investment for board testers, remember that the final goal is the reduction or avoidance of board tests. The testers first used to detect manufacturing faults will then be transferred to the manufacturing line. The self test of intrachip logic leaves dynamic functional testing of interchip logic to the board test.

Failure diagnosis

Board testing is not a go–no go test; all faulty boards have to be diagnosed. A general overview of tests and of failure diagnosis is shown in Fig. 6.9 in conjunction with Fig. 6.10(a) which refers to board testing. The first step is to decide whether the failure is caused by bad manufacturing or by a bad component.

In the latter case, the incoming inspection procedure has to be repeated on the questionable component. If this reinspection confirms the failure then the component is subjected to a detailed failure analysis procedure as described in section 6.9. Infant mortality, ESD sensitivity, or gaps in incoming inspection with respect to vendors' AOQ are frequent causes. Suitable countermeasures can be derived from the result of this analysis.

If the failure is not confirmed, then the reason may be due to an application failure, a sporadic failure or a flaw in the test board.

- An **application failure** occurs when the component is not used in accordance with its specified properties. The designer of an electronic device relied on unspecified parameters subject to sample or lot variations. In this case, the design group has to initiate corrective logic changes.
- Possible reasons for a **sporadic failure** are a synchronizing logic with insufficient waiting time, which is also an application failure, or bad bonding with intermittent lift-offs of the bond wire. The latter can be confirmed by technological analysis.

If the board failure is caused by bad manufacturing (e.g. bad soldering), then feedback to the manufacturing engineers will improve the manufacturing process and product control. Opens are frequent on SMT boards. An open occurs when the lead of an SMD does not sit correctly on its pad or when the coplanarity is more than is tolerable. For through-hole boards, shorts are the biggest problem. A further source of manufacturing failure comes from last-minute changes in board wiring, often carried out manually using discrete wires with insufficient documentation. These failures are completely unnecessary and can easily be avoided by more careful design and more consistent manufacturing.

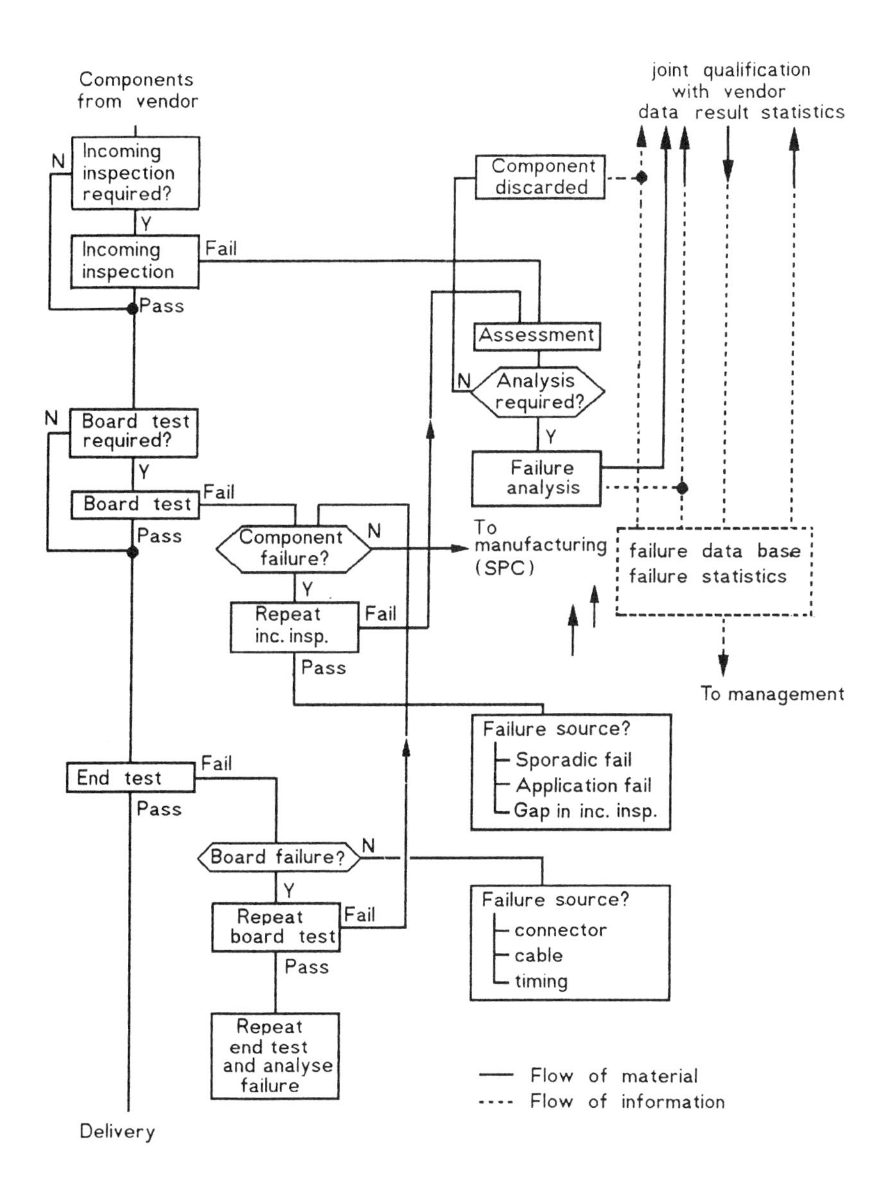

Figure 6.9 Corrective loops. All faulty parts are diagnosed and analysed. The results are fed back to initiate corrective actions.

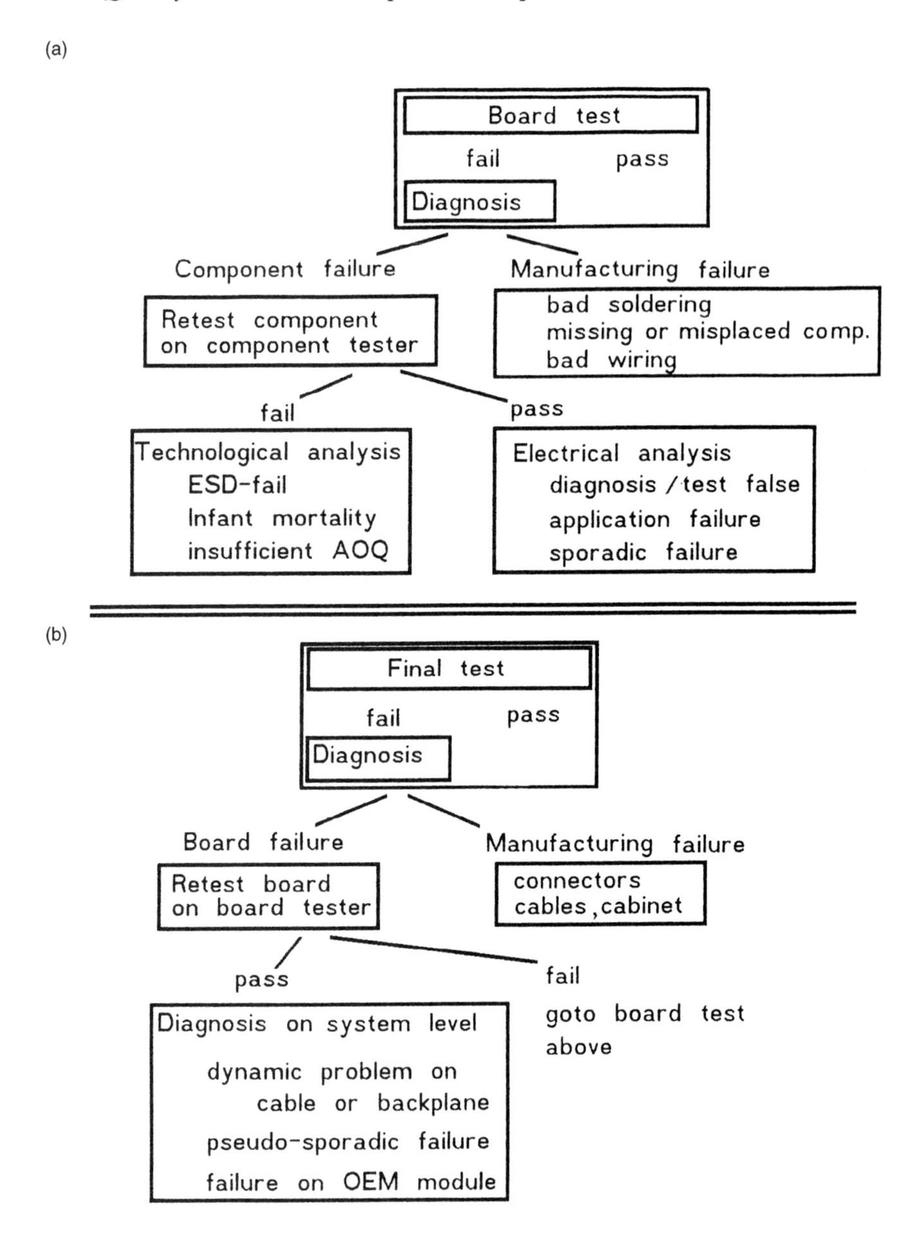

Figure 6.10 Failure diagnosis procedure, cost optimal and effective: (a) board test; and (b) end test.

Documentation

Constant control of quality goals, achieved quality and costs is necessary to make improvement possible. Before diagnosis, all faulty boards are entered into the tracking system described in section 6.2. The results of the failure analysis are added later. This enables the quality engineers to observe the success of the corrective measures proposed above. It offers them a chance of reducing the amount of board testing as soon as the failure rate of the board test and final test declines. Finally, the board test can be discontinued completely and the boards assembled untested in the cabinet. It has to be emphasized, however, that the board test must not be reduced without carefully watching the results of the final test. It is always less expensive to remove board failures at board test.

6.5.3 Design for testability

Design for testability is a very effective measure in reducing test time and test cost for both board and final tests. It has to become an integral part of design and it has to be considered from the beginning or else it may increase manufacturing cost and reduce performance. It is up to the quality manager to insist that the design group observes testability and to find a compromise between the diverging interests. His or her test experience will help to achieve an optimal cost and performance trade-off in the meeting at the end of the planning phase (section 2.3).

Design for testability may be accomplished by:

- using LSSD (level sensitive scan design) within asics;
- providing complete boundary scan on board;
- using components with built-in self test;
- improving accessibility of test points for the needles of pretesters; and
- providing self test on board.

6.5.4 Future trends

Board testing is changing fast at the moment. Due to extreme miniaturization and utilization, the board quality will possibly decrease. On the other hand, there is the necessity to save costs. When incoming inspection has been reduced by ship-to-stock contracts, then the board test becomes the next item on the agenda. Though no trend is apparent, the following can be considered possible:

- a shift from process-related tests like ICT or boundary scan to function-related tests like BIST as shown in Fig. 6.11 – ICTs will possibly migrate to the manufacturing line;
- fully automatic testing, using the intelligence of the logic on board to perform self tests; and
- integrating board testing into the total test strategy – migration of test preparations to final test and service.

A proposal for such a test method is shown in Fig. 6.12. An additional special test processor is integrated on board. It will initiate built-in self tests on all asics and evaluate the test results. It will also perform a boundary scan on RAMs and on bus drivers. A dynamic functional self test of the board can be performed under the control of the test processor. So the board performs a complete test of its own

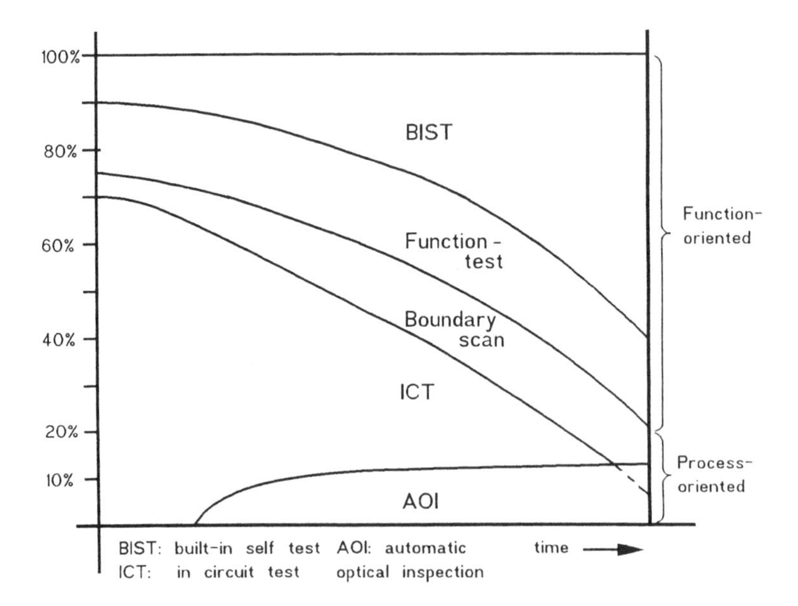

Figure 6.11 Future trend of board testing. Function-oriented tests will replace process-oriented tests. Automatic optical inspection (AOI) for manufacturing faults and built-in self test (BIST) for failing components will remain.

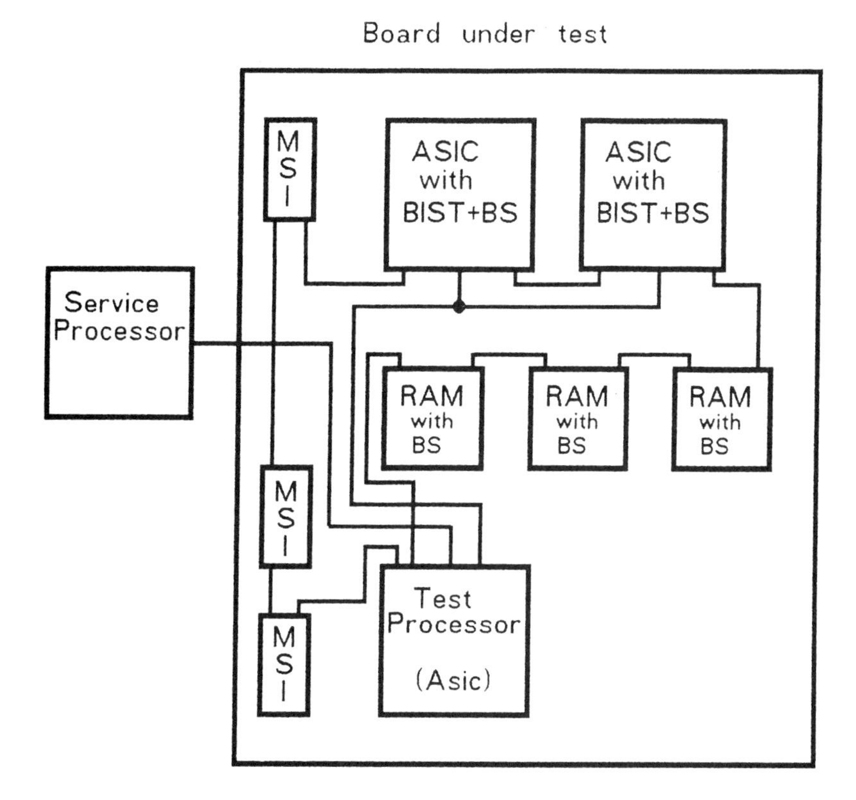

Figure 6.12 Design for test. Proposal for an economically testable board design; each board tests itself. (BIST = built-in self test; BS = boundary scan.)

and no expensive board tester is required. A cheap service processor outside will be the user interface for starting the test sequence, taking over and displaying the result. An additional LED can be used to indicate go–no go. This test arrangement can be of use for both the final test and service.

6.6 MODULE TEST

Large electronic devices may consist of several modules, each module having a clearly defined function, e.g. the RAM module, input/output processor module and hard disk module. These modules are functionally pretested before they are connected to the complete device, according to the rule that it is best to find a failure as early as possible. From this, a sequence of tests is established: board test, module test and final test. But it depends on the complexity of the device whether the module test is really necessary and cost saving. Today, highly integrated parts shrink many modules to a few boards or to one board only, so a separate module test between board test and final test may cost more than it saves. Sometimes, module testing is performed only because nobody thinks about discontinuing it.

6.7 FINAL OR END TEST

6.7.1 Test procedure

Final test methods depend more on the spectrum of products manufactured than is the case for incoming and board tests. For the following section, medium-sized mainframes or large-sized workstations have been chosen to demonstrate the problems of final testing. Small PCs would have been less informative and large mainframes too specialized.

The final test is performed on the complete unit consisting of a rack or cabinet, boards, power supply, internal peripheral devices and cabling.

It is assumed that all these components were submitted to one or more preceding test steps, either 100% or on a sample basis, in order to detect each failure at the earliest stage of production. It is not the objective of the final test to find failures, but to ascertain that the product is fault free. It has to ensure that the product delivered to the customer has the guaranteed quality in terms of dpm. Therefore all functions of the product are tested, hardware as well as software. The final test is usually performed in several main test sections each consisting of several test subsections. The main sections are listed below (Fig. 6.13):

1. **Pretest**
    - Control of serial number and configuration – this is important for systems with internal bus structures which allow many different configurations to be delivered according to customers wishes
    - Tuning of clock oscillators and timers
    - Test of switches and coders

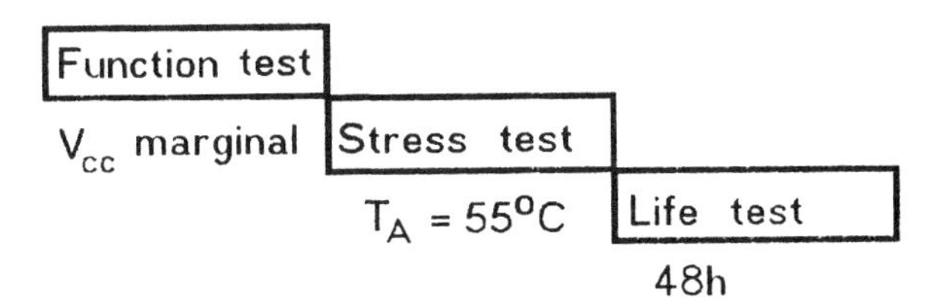

Figure 6.13 The three main sections of the end test.

- Preloading of test software
- Connection to a service processor
- Contact test of board connectors to back plane.

2. **Main test**
   - Functional test (design checks) static and dynamic
   - Quality control programs (marginal supply voltage, frequency and temperature)
   - Test of internal peripheral devices (e.g. HD)
   - Test of peripheral controllers with simulators substituting the external peripheral devices
   - Software tests (operating system and customer programs)
3. **Life test**
   Depends strongly on the quality of the components and of their burn-in – a 48-hour life test is common to keep warranty costs low.

Traditionally all these tests are performed manually by experts. This is very cost intensive and time consuming. Three-quarters of all test costs are spent on the final test after the incoming inspection has been reduced due to ship-to-stock procedures and the board test has been optimized by a reduction of manufacturing failures due to successful feedback. So, the remaining major task is to reduce the cost of final testing too.

This can best be done by the following procedure:

- refer all quality costs to the products by which they are caused;
- sum up all quality costs for each product separately;
- set prospective goals for cost reduction;
- look for failure clusters, and set up a hypothesis for their cause;
- plan measures to avoid these failures, and set time limits for their realization;

- calculate the potential cost savings; and
- implement the planned measures.

After the time limit has elapsed, a control procedure should follow.

- Did the measures show the intended effect?
- Are the quality costs in accordance with the goals?
- Are new failure clusters occurring? Which additional measures are necessary?

To plan measures to reduce the costs of final test, it is necessary to look again at Figs 6.9 and 6.10(b). The test flow and failure diagnosis of the final test are shown in these figures. A great deal of the failures may still be board related, either component failures or failures caused during manufacturing. The faulty boards are removed and subjected to a second board test. If the failure is confirmed then the same failure analysis will be performed as for board tests (section 6.5.2) with all its consequences. If not, then the board has to be put back into use in the system for further manual analysis. If this analysis points to an incomplete board test, then the gap in that test has to be filled. Other failures which at first glance are attributed to bad boards, explained later by uncalculated back plane or cable delays, are design problems. Further possible reasons for failure at final tests are bad manufacturing of the back plane, connectors and cables.

From the above overview, proposed measures to reduce costs can be derived. You can reduce the number of failures per system by a further improvement of the quality of the parts (boards, cables, OEM products):

- Board failures should be detected at board test. If a board failure has been detected at final test then a failure analysis is essential. Feedback of results should produce more stringent board tests and/or improvement of the board test procedure (section 6.7.2).
- The same is true for pretests of cables, connectors and back planes. All failures on these parts must be detected before final test by a suitable test method.
- Failures of OEM modules should be detected on incoming inspection. OEM modules lie between components and boards. They are bought like a component; however, their logic is more complex than a board. Therefore, in most cases, they are subjected to a substitution test by putting them into a working system. This may be done on a 100% basis at first, and on a sample basis later on. The vendor is usually willing to replace

faulty modules at no cost and reacts only on failure clusters. At first glance, this seems to be acceptable but gives little hope of reducing the test cost. Co-operation with the vendor to improve quality, as has been done successfully at the component level, is recommended. The goal should be to achieve a failure rate of less than 2% at final test.

The failures that remain include:

- board failures caused by infant mortality or by electrical overstress of components after passing the board test, during assembly, or at final test;
- timing problems between boards (in the back plane and cables) – these failures are sometimes of a pseudo random nature;
- manufacturing faults caused during assembly of the cabinet; and
- sporadic failures.

It may need weeks to detect sporadic failures and they often become severe problems at customer's site (section 5.11.2 for example). Main reasons for sporadic failures at final test are:

- synchronizing failures – the waiting time when synchronizing asynchronous data coming from different units may be too short;
- pseudo-sporadic failures – failures which occur seldom and are seemingly uncorrelated to the applied test sequence, but this is only because they are dependent upon a rare combination of logic data and environmental conditions; and
- EMI-caused failures in spatial distributed systems.

To control EMI two effects have to be distinguished: the noise radiated by electronic systems which may disturb other people's systems; and the noise penetrating into an electronic system from outside.

Effective shielding is necessary to reduce EMI-caused noise. Shielding effectiveness (the ratio of a noise-generating electrodynamic field without a shield and with the shield in place) is primarily the responsibility of design and engineering and not of testing. Whereas most designers are aware of capacitively or inductively coupled noise and avoid wide openings in the cabinets, external ground noise is often neglected. Considerable external ground noise can be generated within an electronic system consisting of several separate units interconnected by cables. It is even worse if these units are scattered in a building with their power supplies connected to separate AC power distributors. Theoretical considerations, explained in Appendix G, require a perfect shield, uninterrupted from cabinet to cabinet.

By means of electromagnetic coupling, a voltage will then be induced in the signal wires inside the cable to compensate for the external noise voltage. This is the point where the final or system test comes into play.

Any discontinuation of the shield (e.g. by a deteriorated or insufficient connection from the cable shield to the connector case or from the connector case to the cabinet) reduces the shield effectiveness considerably by several orders of magnitude. This may cause sporadic failures of the system originating from switching high currents nearby (e.g. motors, high voltage switches, fluorescent lamps, etc.). Therefore, in case of EMI problems at final or system test in distributed systems, checking the ground connection between cable shield and cabinet for a low impedance first is recommended (Fig. 6.14).

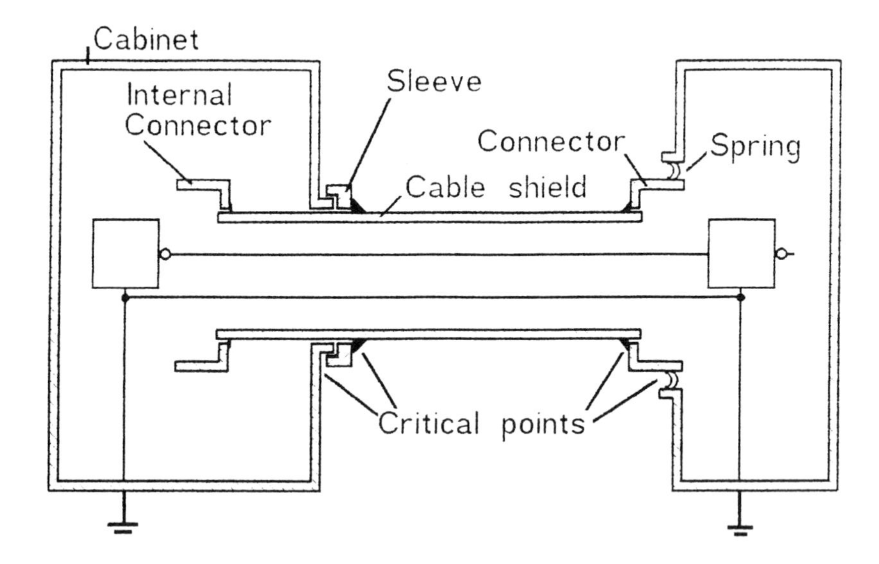

Figure 6.14 How to connect two cables without EMI. A continuous and uninterrupted shield connection with low transfer impedance between the cabinets is important. Connectors are a weak point.

You can also save on the time and cost of final testing by:

- migration of board tests like built-in self test to the final test;
- optimal scheduling of test steps;
- test automation by a test handler; and
- the use of an expert system.

All these measures are explained in detail in the following sections.

6.7.2 Detect board failures at board test, not at final test

The situation before and after an improvement of board tests is shown in Fig. 6.15. About 10% of the total manufacturing cost is spent on quality assurance in the production area. Not included are the cost of incoming inspection (around 0.5%) and of evaluation. Final testing accounts for 78% of these quality costs. This is due to the high rate of manual testing performed by experts. Many of these costs are spent on failures which should have been detected by the board test. With failure analysis and feedback, it was possible to improve board testing to the extent that 95% of manufacturing and 80% of component failures were detected and eliminated there. This increases board test costs, of course, but the cost of final testing are reduced far more. Detecting a board failure manually at final test, as reported by one factory, costs about 32 times more than at a fully automated board test (800 US$ against 25 US$). Even assuming that board test costs doubled through installing additional test software, a reduction of quality cost to around 6% was achieved by this single measure.

This depends on the complexity of the product and on the status of the final test. Another factory stated a factor of 10 (150 US$ against 15 US$).

Final test	
component board failures: 20% of 35%	= 7%
manufact. board failures: 5% of 30%	= 1.5%
OEM design and rack failures	= 35%
Total final test	= 43.5% of previous cost

Quality cost	
final test: 43.5% of 78%	= 34%
add to board test: 56.5% of 78%/32	= 1.37%
add test software	= 1.37%

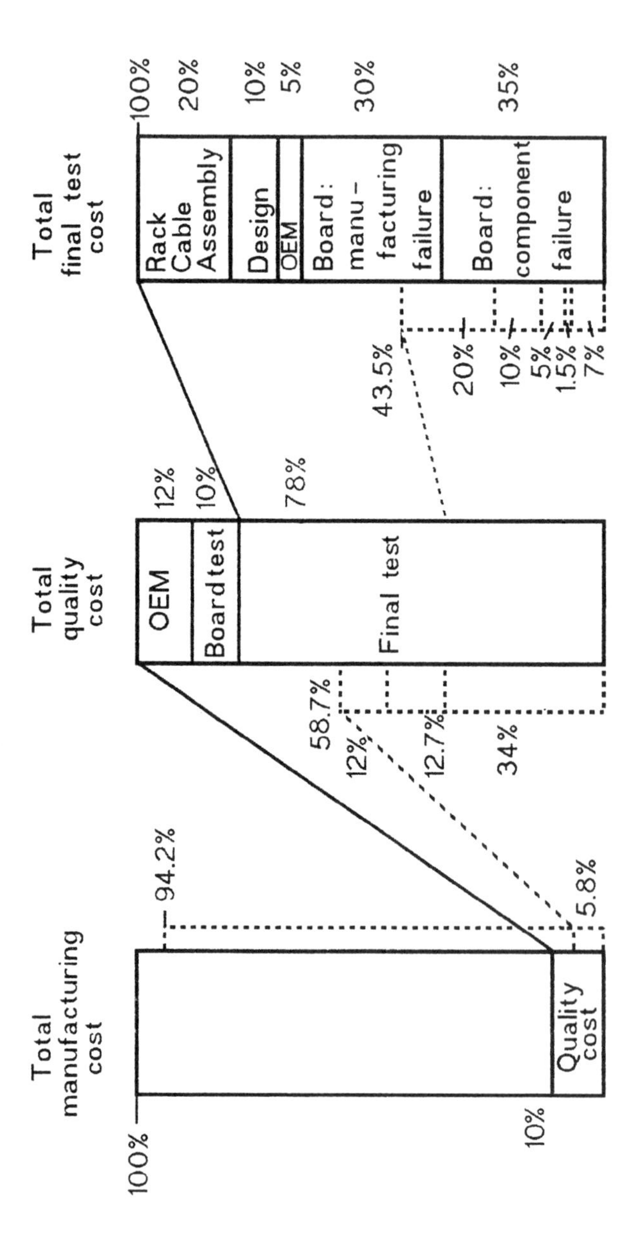

Figure 6.15 The final test must not detect board failures. It is far more economic to screen out all board failures at board test.

resulting board test	= 12.7%
OEM test	= 12%
Total quality cost	= 58.7% of previous cost
	= 5.9% of manufact. cost

A similar procedure has to be adopted with OEM components.

6.7.3 Optimal scheduling of subtests

A first step in reducing test time is to arrange the subtests in such a way that those showing a high failure rate are performed first. The success of this measure can be derived from Fig. 6.16.

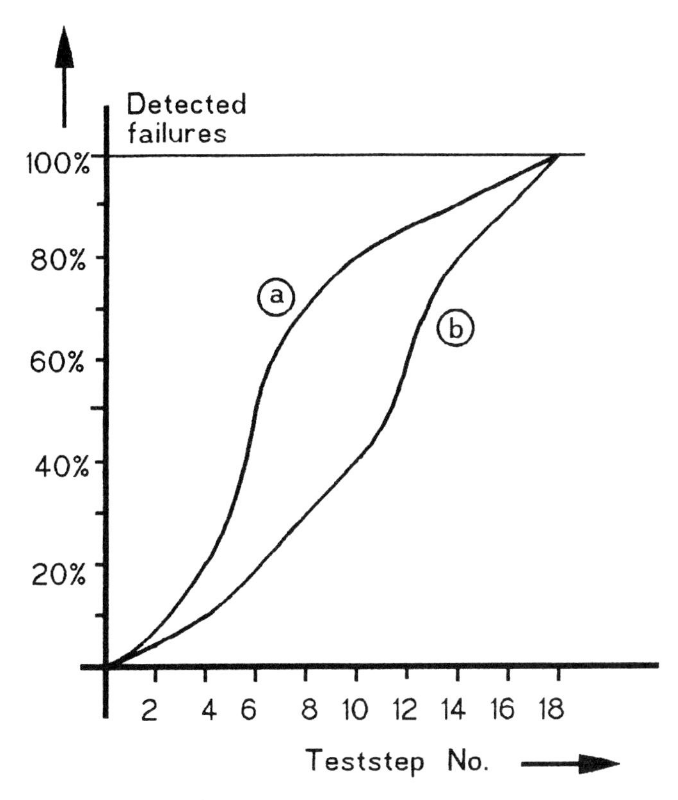

Figure 6.16 Detect failure as early as possible. This principle is also valid inside a test. Schedule subtests in an optimal sequence: (a) subtests with a high detection probability are performed first; and (b) random sequence.

6.7.4 Test handler

The test handler is a software package which automatically runs the sequence of tests to be performed during final testing. It can:

- recognize, check and control the configuration automatically;
- load the correct test software from a general database;
- start and stop the test software automatically for each test step;
- automatically schedule all tests which require manual operations (e.g. insertion into a temperature chamber) to a time when the tester is being operated (e.g. daytime);
- react to failures, retest to confirm the failure and file a failure note; and
- generate a print-out of a failure protocol.

This automation achieves considerable cost reductions. The cost of running the final test program has been reduced by 77% (Table 6.6).

Table 6.6 Cost benefits using a test handler

Test section	*Relative cost*		*Degree of automation*
	% without handler	*% with handler*	*%*
Pretest	7	6	14
Design check	18	2	89
I/O check	26	8.5	67
Function test	45	4.5	90
Software test	4	2.5	37
Sum	100	23.5	76.5

Base for all numbers is the cost of running the final test program without an automatic test handler. The first column shows the relative costs for each test section and the second column the same using test handler software. The third column shows the achieved degree of automation. The cost of failure search and diagnosis are not included.

6.7.5 Use of an expert system

The problem

Conventional test methods for performing end tests require all test engineers to have a thorough knowledge of the test object. The high rate of innovation means a rapid change in the subjects to be tested and in their complexity. Much has been done to facilitate locating failures with hardware and software assistance. The use of a service processor (SVP) and of level scan technology are examples of this. Nevertheless, it seems impossible that expert knowledge of all possible failures on all actual systems will be available at a tolerable expense in the future. At the same time, demands are increasing to reduce time and increase the quality of failure diagnosis.

The solution

The solution to this problem is an expert system which will allow analysis of each failure down to the component level. The purpose of this expert system is to give recommendations to the test engineer on how to proceed to eliminate the failure. This system will also be available to maintenance and service.

The procedure

On detecting a failure at final test, the test software starts a failure analysis routine, preferably assisted by a test handler. Usually, the failing test step is repeated to confirm the failure. Independent of this analysis, a failure note is laid down in a failure file. The note comprises all data concerned, e.g. the situation of the hardware at the time of failure. This failure log out is then analysed by the expert system and the result printed on screen or paper. This result contains all important information on which parts of the test object might be involved and allocates a priority. Figure 6.17 shows a reprint of such an expert system recommendation.

As shown in Fig. 6.9, the corrective loop comes into action. The failing boards return to board test according to their priority. If the board test confirms the failure, then the board is repaired; if not, then the component to which the expert system points is changed. Where there is no success, the conventional method of end testing comes into action, i.e. humans search for the failure and the expert system

```
Logout  : X0081   Date: 12.05.92      Time: 12:38:17
CPUID   : 0400600708993               Type: PCD22
-------------------------------------------------------
Unit     : UC3 RAM
Errorbit :  ERW78 IERR314
Prio1
  Board : D1008    M12
    Chip : UPC3     J70
Prio2
  Board : D1169   M13
    Chip : ERRC5  J39
Prio3
  Plug
-------------------------------------------------------
Remark : If the failure migrates to RAMB
when switching off RAMA then Prio2
becomes Prio1
```

Figure 6.17 Print-out of an expert system protocol, pointing to boards and components most likely to be faulty.

database is completed. This method extends the knowledge base of the expert system.

Knowledge base

The knowledge base contains not only the facts but also the rules as if–then commands. The quality of an expert system depends on filling the database. It is absolutely necessary that all experts concerned make their knowledge available. This requires a high degree of technical knowledge, and high motivation. Special knowledge engineers extract experts' knowledge and formalize it so that the inference component of the system can be applied.

It is most important for the expert system to be ready for use when the first systems arrive at final test. To achieve this, build up the knowledge system in parallel with the design. The acquisition component of the system has to be an easy-to-use synoptic data input facility. During the final test, it is the knowledge engineers' task to extend and diversify the knowledge base.

Results

While testing a medium-sized computer system, the following results were achieved:

- running one analysis on the expert system took a maximum of five minutes;
- up to eight hours of test time were saved by using the expert system; and
- the hit rate for priority 1 was 75% for defective boards and about 60% for defective components.

6.7.6 Achieved cost reduction

Figure 6.15 shows the cost reduction of final testing achieved by shifting test efforts to pretests. Figure 6.18 shows the improvement possible if all the measures proposed in this section (6.7) are put in place.

In Fig. 6.18(a), the reduction on the total number of failures and in the number of faulty boards per device is shown. This reduction is due to successful feedback to manufacturing and by improving pretests. Figure 6.18(b) shows the reduction of the mean test time of the end test per device; this was mainly the result of the introduction of a test handler and an expert system. In Fig. 6.18(c) we see the total costs of end testing drop from about 10% to less than 4% of the production costs. All these are mean values derived from several device types and production sites. Therefore, only relative values can be given and some time adjustments had to be made to compensate for the different rates of progress.

6.8 SERVICE AND MAINTENANCE

The quality of a product is completely determined by the success of quality assurance in the production phase and cannot be improved by service. Table 6.7 shows how the MTBF of products increased in the last decade. The cost of maintenance is directly related to the number of service calls. Therefore, better quality of the products, expressed by higher MTBF, is the key to cost reduction of service, because the competition set the standards for warranty time and price limits for service contracts.

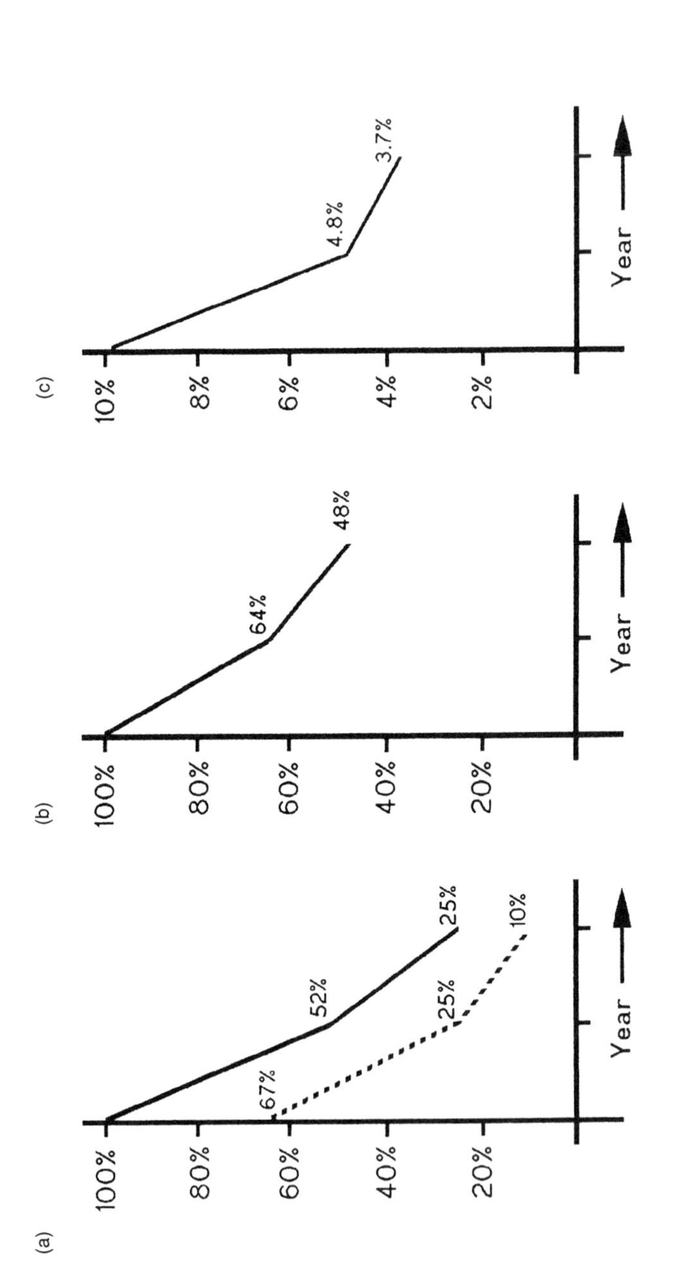

Figure 6.18 Improvements in end testing: decrease in (a) the total numbers of failures and of board failures per device; (b) the test time per device; and (c) the cost per device as a percentage of the manufacturing costs.

Table 6.7 Improvement of quality in the last decade

Product	*MTBF* (h)*		*Factor*
	1983	*1993*	
Electromechanical device (printer)	300	3000	10
Medium mainframe, workstation	3000	45 000	15
Small computer, peripheral	4000	160 000	40

* The mean time between failures (MTBF) has increased more than ten times in the last decade.

The main problems of today are:

- fewer connections between customer and vendor – reduced barriers to the use of foreign service because of open systems and down-sizing, and the customer requesting single source service for all products;
- more globalization and disintegration of the market; and
- rising product quality and warranty time for new products.

Your answer to these problems should be to increase customer satisfaction and to reduce costs by increasing efficiency.

Measures to achieve this are:

- decentralization of service organization;
- lean headquarters responsible for strategic decisions only;
- single-stage service organization, with support centres near the customer;
- avoidance of unnecessary service actions by analysis of all service calls at the support centre;
- migration of test tools from final test to service; and
- integration of service requirements into product design (Fig. 6.19, design to service).

There are two service requirements:

1. Increased MTBSC (mean time between service call), with a goal of MTBSC = MTBF to be achieved by:

    - simple architecture
    - increased MTBF by greater integration and better cooling
    - installation by the customer (plug and play)

- replacement of wear-out parts by the customer (lamps, batteries)
- bring-in by customer (the customer bringing the equiment to the service centre)
- software service by hotline.

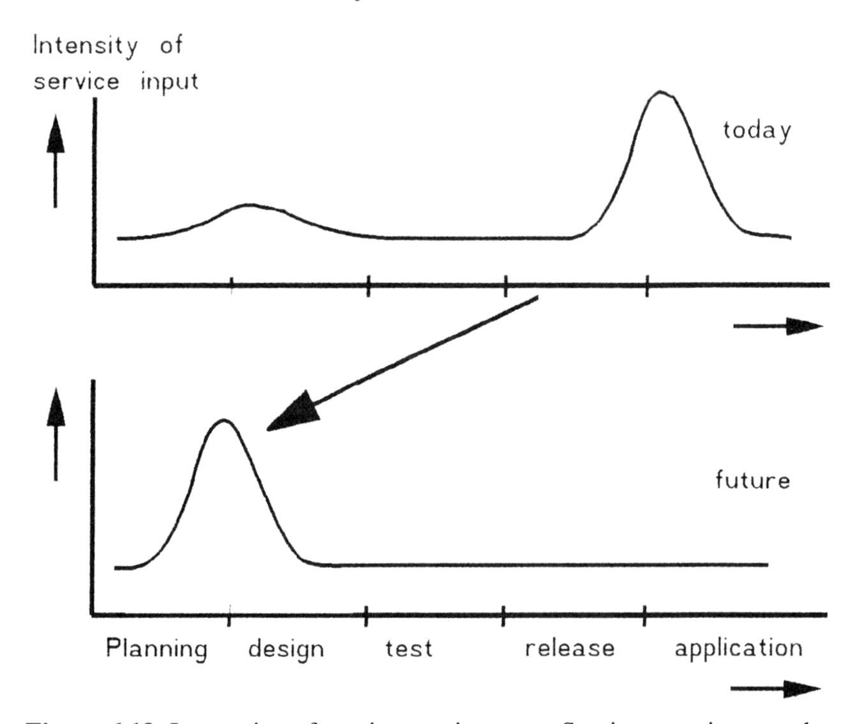

Figure 6.19 Integration of service requirements. Service experience and customer complaints are considered early in a new product design. Job rotation of service engineers to increase their consciousness of design problems and to gain acceptance of their problems by design engineers is an effective measure.

2. Reduced MTTR (mean time to repair), with a goal of MTTR less than 5 h, decreasing to 1 – 2 h) to be achieved by:

   - improved diagnosis software
   - use of hardware tools such as on-board diagnosis (boundary scan), and on-chip diagnosis (BIST)
   - readable service messages
   - teleservice
   - assistance by an expert system.

Cost considerations

Warranty costs are part of the production costs and not the cost of quality assurance. Good outgoing quality, however, reduces warranty costs, so quality assurance is involved in service too. One example showed that an increase of the MTBF by a factor of two reduced the warranty costs of a product by 42%. The entire life cycle of the product has to be considered when calculating production costs.

6.9 FAILURE ANALYSIS

6.9.1 Failure documentation

To perform failure analysis, it is essential to document each component failure detected by incoming inspection or by a test during the manufacturing process. This documentation has to include all details, part number, vendor and date code, tester type, the test engineer and the symptoms observed, the location on the board, the faulty pin, frequency, ambient temperature and supply voltage if applicable. All this data has to be fed into the tracking system database mentioned in section 6.3. The faulty part itself has to be stored in a manner which allows it to be identified later. One way to do this is to put them into bags whose inscriptions are generated automatically by the tracking system (Fig. 6.20). This system enables statistics to be made showing the failure rate dependent on vendor, family, type, lot, board type or manufacturing step.

6.9.2 Execution

The documented defective parts should then be subjected to a detailed failure analysis. The assignment of a failure to its failure-creating mechanism is a necessary requirement for a fast and effective reaction. Successful failure analysis requires a consistent concept. First, a barrier has to be set up. For reasons of cost-effectiveness, only parts which have a failure rate exceeding a certain level should be analysed. This level depends on the overall quality of the components and is variable. When the failure rate is high, then only so-called 'asparagus types' are analysed: component types which have abnormally high failure rates. As soon as the quality improves, the barrier is lowered to improve quality further. This adjustment of the barrier and the decision to do a failure analysis is an example of a quality action.

From Board/Endtest

Failure description

defect pin/lead:(..)

Category of failure:

Temperature: (......) Dynamic: (......)
Supply nom.: (......) Static: (......)
Supply max.: (......) Level: (......)
Supply min.: (......) After: (......)h
Sporadic: (......)
Testprogram No.:(...)
Teststep/line No.:(...)
Serial No. of Device:(...)

From Component Retest

Tester:(.......................) BIN:(..........)
Failure confirmed? yes(......) no(......)

From Board/Endtest

Failure retested? yes(......) no(......)
Failure confirmed? yes(......) no(......)
Remark: (...)

Figure 6.20 Bags to store components. The test history print-out is made by the tracking system.

The electronic systems manufacturer may decide to let the failure analysis be done by a specialized test laboratory or by the vendor, for reasons of cost effectiveness. Nevertheless, for a qualified manufacturer, some kind of failure analysis is essential. Even if users have decided not to perform the reverse engineering process, as described on p. 212, alone, they should be able to assess the results. This needs some knowledge and experience of how to interpret the chip photographs made by the analysts. One recommended method is to set up a catalogue of sample chip photographs with annotations (Burggraf, 1984).

6.9.3 Process of analysis

An overview of the process of failure analysis is shown in Fig. 6.21. The following actions have to be performed:

- **Electrical verification and localization of the failures**
  It is nearly impossible to analyse a failure without a detailed and exact description of the failure effect. If a package is opened without a preceding failure localization, then one will find many suspicious spots but very likely overlook the real cause of the failure. Special software is often necessary for this failure localization, if the circuits are highly integrated. The software should detect the category of the failure and the region to be searched. For dynamic RAMs, for example, the logical address has to be transformed to the physical one.
- **X-ray analysis**
  Transmission radiography produces an enlarged shadow image of the material contrast of the DUT on a detection screen. The situation of the chip inside a plastic package can be seen and rough defects of bond wires detected. This is usually a pretest and may save time.
- **Opening the package**
  This is fairly easy for ceramic packages but for plastic packages straightforward procedures are also available. First grind the upper side of the package and then etch the whole package with NHO_3 until the upper side of the chip becomes visible (the chip being carried by the lower half of the package). Selective etching is seldom necessary.
- **Optical inspection of the part by microscope**
  Objectives with a numerical aperture of 0.95 are available which results in a resolution of 0.6 μm. About 70–80% of all failures can be detected optically; for the remaining 20–30% a SEM (scanning electronic microscope) has to be used. Hidden layers can be observed to some extent because of the transparency of the dielectrical layers.
- **SEM analysis**
  If the optical inspection is insufficient, then a SEM has to be used. A SEM photograph is excellent for analysis and documentation

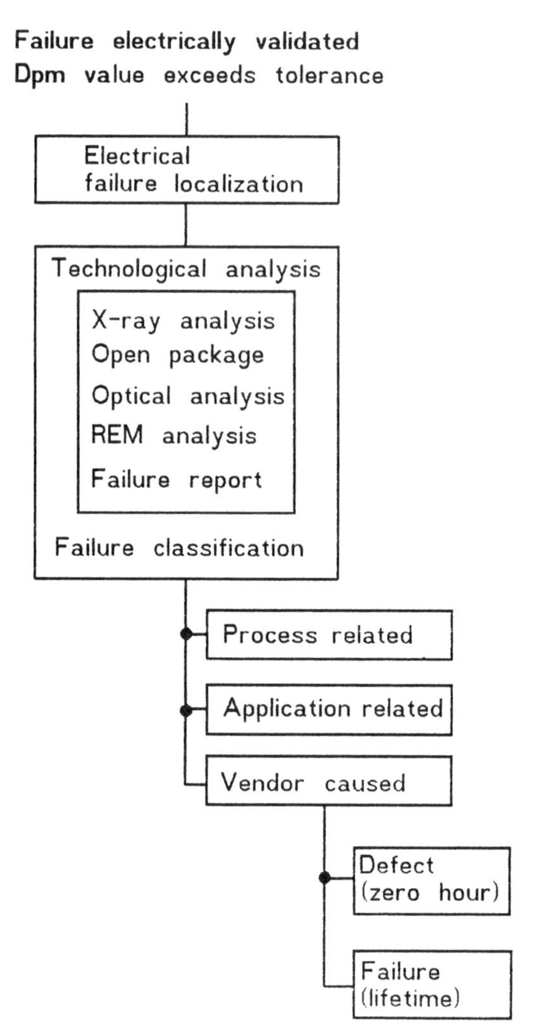

Figure 6.21 Failure analysis procedure.

because of the large depth of field and the high resolution of 2 nm. Therefore a SEM has become a standard instrument for failure analysis. A magnification of ×1000 is appropriate (1 cm ~ 10 μm).

- **Reverse engineering**
  If the failure is hidden by upper non-transparent metal layers, then these layers have to be removed by reverse engineering. For new technologies, ask your vendor for a recommendation for the reverse engineering process.

- **Test by DFI**
  Some DFI (dynamic fault image) testers automatically compare the images of good and bad parts to detect the fault location. According to the experience of users, these testers are fine for manual analysis but today really fully automatic work can only be done on very simple or very regular circuits.
- **Use of STEM or AES**
  STEM (scanning transmission electronic microscope) or AES (auger electronic spectroscopy) are used to identify materials.

Other instruments are used by manufacturers of components and specialized test laboratories.

6.9.4 Interpretation of results

Most failures are located on interconnection lines on a chip; very few (less than 1%) are on the diffusion layers of the transistors. The most likely problems are displaced masks. Figure 6.22 shows a REM photo of a 256 kbit DRAM with some openings (insufficient overlap between poly 1 and poly 2) caused by bad lithography. Figure 6.22(a) shows how difficult it is to find this kind of failure without a preceding failure location.

A short circuit between two metal lines caused by a similar manufacturing defect is shown in Figs 6.23(a) and 6.23(b).

Figure 6.23(d) shows another short circuit, but caused by an iron particle. Particles are another important reason for manufacturing-related failures.

All the preceding failures are created by manufacturing under the responsibility of the vendor. The following failures are originated by the user. Figure 6.23(c) shows a defective PAL circuit. The fuses are insufficiently blown. Inadequate ESD handling is the reason for the failure near the bond pad of Fig. 6.24(a) distinguishable by the spark gap. Both are failures which are related to the user's manufacturing process.

Figure 6.24(b) shows two failures caused by an excessive current through a bond wire. Beside the left bond pad, purple plague can be seen; the right bond pad shows a brittle fracture caused by recrystallization. These may be caused by bad handling – more likely by an inadequately designed application. They are application-related failures.

These examples of SEM photographs and their interpretation may be bewildering for an inexperienced quality engineer. Even if he or she does not have the time to become an expert, with a little training he or she will have some feeling for the matter and will no longer be fully dependent on the explanations of experts.

The continuous reduction of failures at incoming inspection by the considerable improvement of component quality is reflected in the amount of failure analysis necessary. This has been confirmed by some quality managers. The number of failures to be analysed has been constantly decreasing in the last few years. But technological analysis is not performed just to analyse defective parts; its second objective is to analyse new components made in a new technology or from a new vendor. A judgement of the technological ability of a new vendor may sometimes save or reduce expensive and time-consuming environmental tests. One manufacturer of electronic systems said that the percentage of evaluation-caused analysis rose from 20% in 1985 to 80% in 1995, whereas the percentage of failure analysis behaved vice versa.

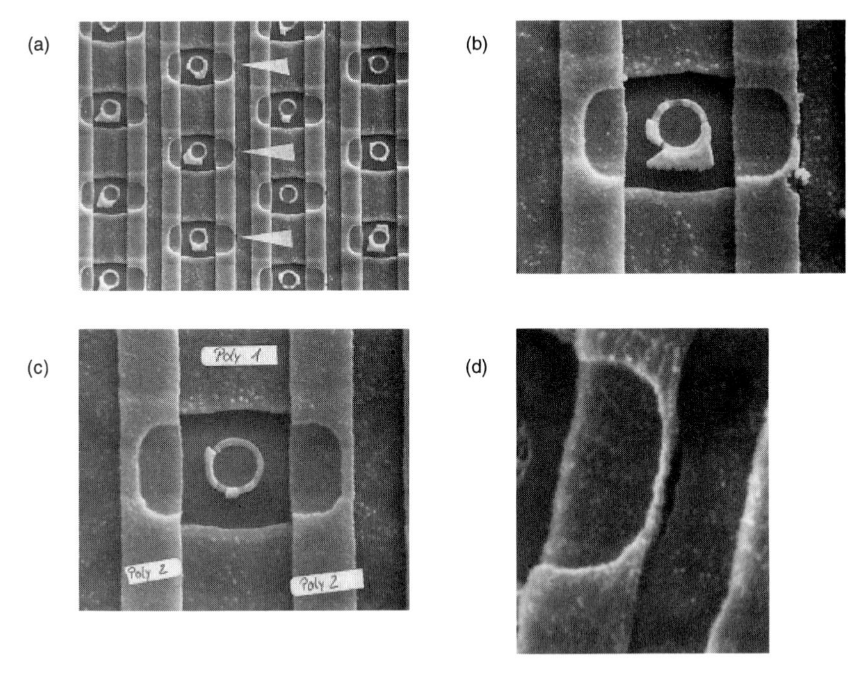

Figure 6.22 SEM photograph of a mask failure: (a) an overview photograph shows how difficult it is to find failures (arrows point to the failing locations); (b) enlargement of faulty areas; (c) enlargement of faultless area; and (d) further enlargement, where the dark spots are the holes created by insufficient overlap between poly 1 and poly 2.

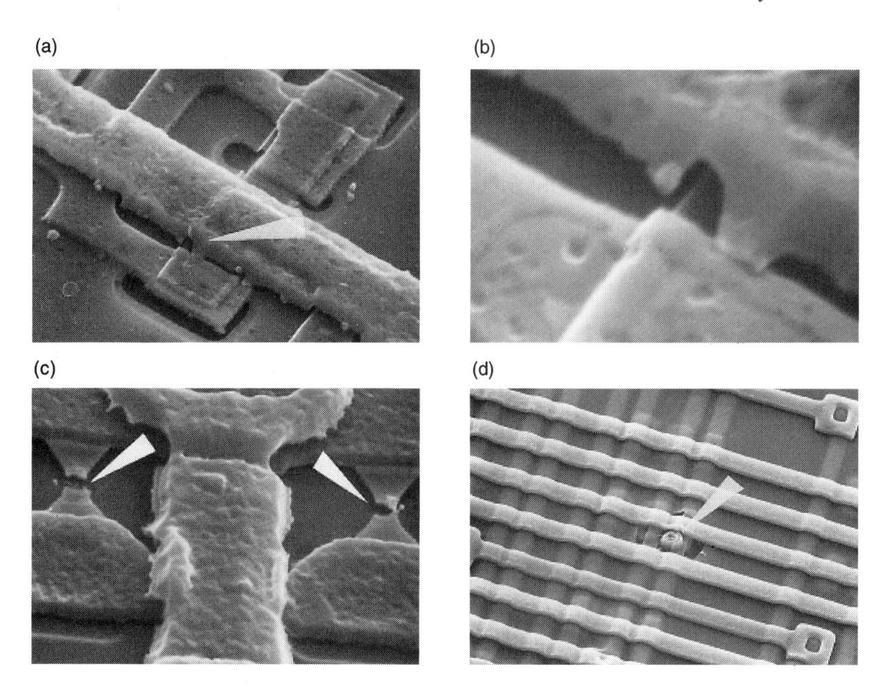

Figure 6.23 More examples of failure analysis: (a) short circuit between ALU 1 and ALU 2; (b) an enlarged photograph shows the details; (c) fuses of a PAL are not blown completely, a frequent PAL defect; and (d) an iron particle causes a short circuit between the two metal layers.

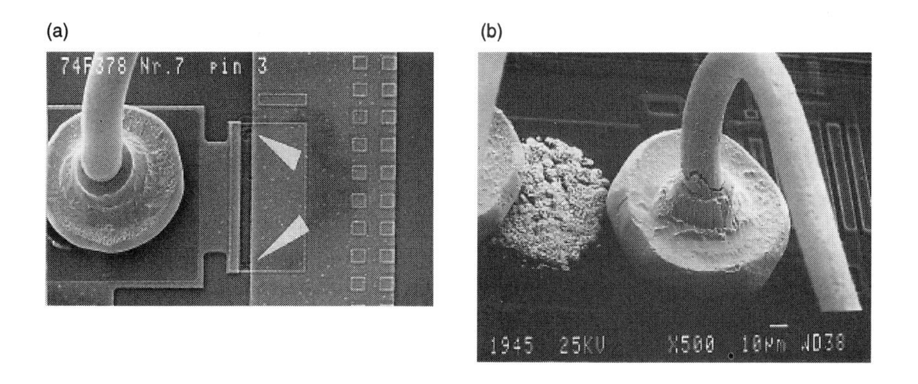

Figure 6.24 Bond failures are easy to detect: (a) sparks caused by ESD leave their traces usually near the bond pad; and (b) two frequent bond failures, purple plague (left) and brittle fracture (right).

6.10 CORRELATION WITH THE VENDOR TO REDUCE FAILURE RATE

Manufacturers of electronic components are eager to receive feedback from their customers about the field quality of their products. The failure rate of modern ICs is so low that life tests at the manufacturer's site will not give much useful data for further improvement. The manufacturer has to rely on actual field data from customers.

A key manufacturer of ICs said that, in general (seen worldwide), only 15% of all components sent back from customers turn out to be really defective. On the other hand, about 50% of the rejects from customers who verify the faults with a rough functional check before sending them back, are justified. And if a customer performs a detailed failure analysis and delivers failure documentation along with the failed samples, then about 85% of the complaints are confirmed by the manufacturer. This helps considerably in detecting weak points in its circuits and in improving products further.

In this case, the vendor will be willing to inform the customer of the results of the failure analysis and the corrective measures being planned. In future, the vendor might decide to perform a failure analysis with the customer and give information of all corrective changes in the products. The customer should also return his failure statistics to the vendor, so the vendor can control the effectiveness of the corrective measures used. Close co-operation and a quick flow of information between vendor and customer, as shown in Fig. 6.25, is essential. This is the first step towards building up trust between the vendor and customer.

Figure 6.26 shows the reduction of the defect rate at incoming inspection and the improved correlation with the vendor. The numbers in this figure only represent a snapshot of a special case but the trend is generally valid. The cost per tested component (Table 6.8) is derived from the equipment cost (Table 6.1). The cost of testing is independent of the defect rate, which means that the cost per failure rises when the quality improves.

The increasing cost per failure leads to the conclusion that considerable savings can only be achieved through the gradual reduction and final cessation of testing. The conditions for this are improved quality and a close relationship of trust. The 100% test can be replaced by sample tests and then, continuously observing the defect rate, the components can be delivered to stock without test.

Table 6.8 Test cost per failure

IC type	*Test cost per IC (US$)*	*Failure rate (dpm)*	*Test cost per failure (US$)*
Processor	0.5	330	1500
ASIC 100kgate	0.6	300	2000
DRAM 4 Mbit	0.25	200	1250
SSI/MSI	0.075	30	2500
SSI S-T-S	0.024	30	800

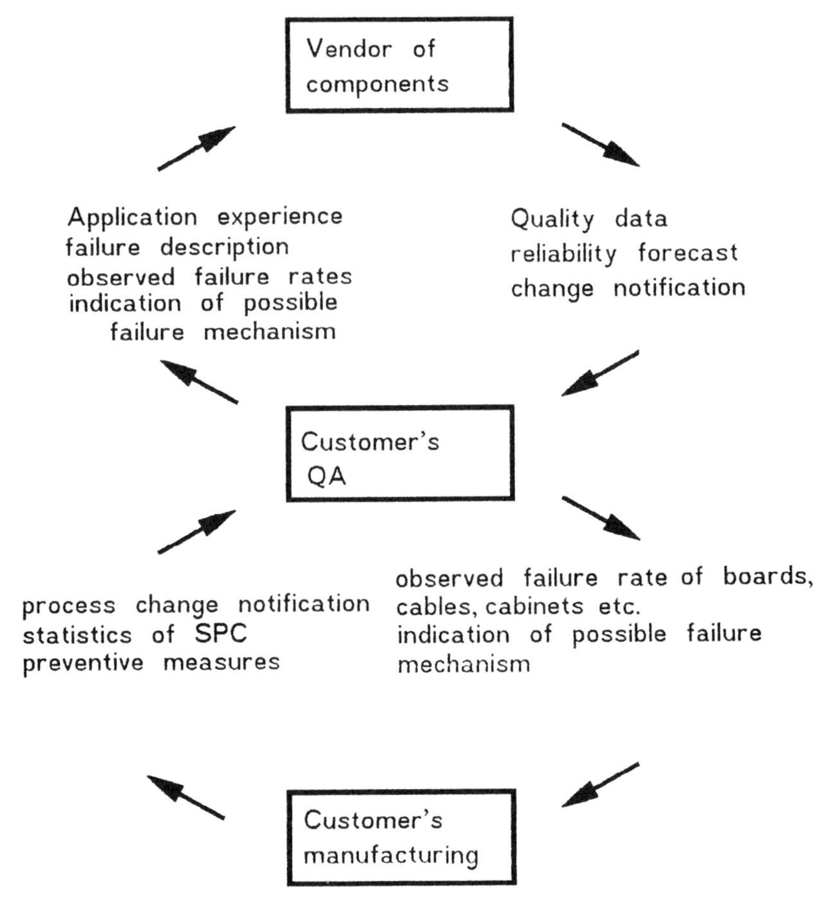

Figure 6.25 The quality assurance management as an information centre. An intense exchange of information between vendor and customer is the basis for mutual trust; the same is true between quality management and manufacturing.

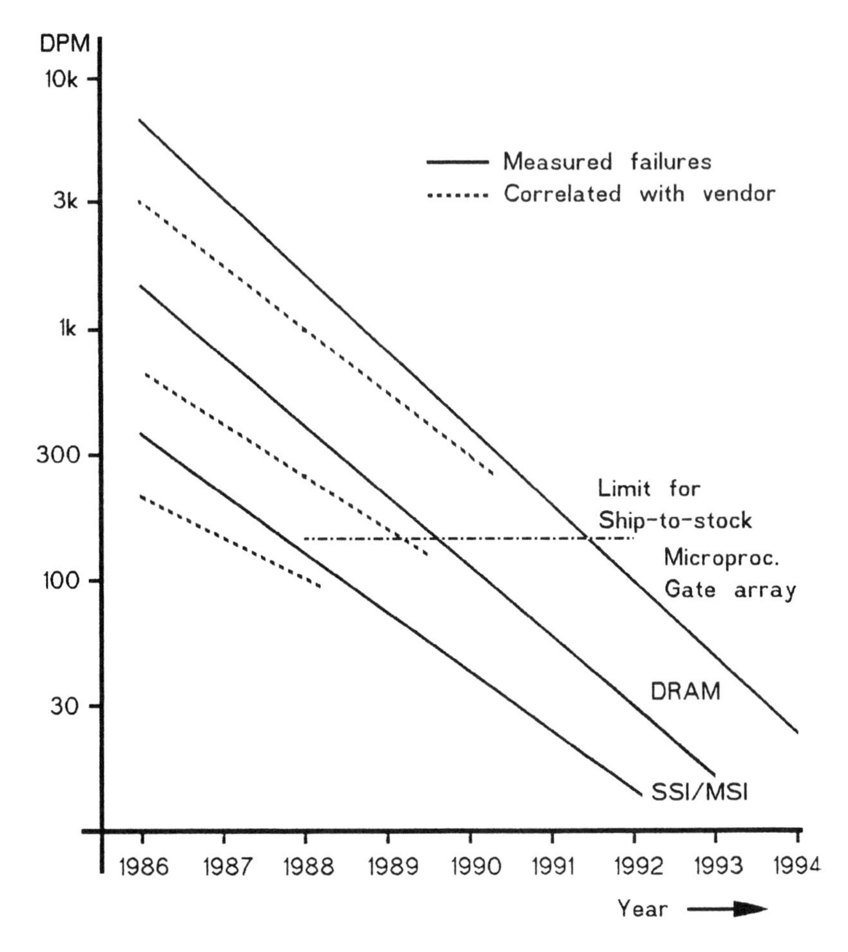

Figure 6.26 Failure rate at incoming inspection and board test. A considerable improvement has been obtained in the last decade by a joint effort of the manufacturer and user of ICs.

This change in incoming inspection philosophy is shown in Fig. 6.27 in conjunction with Fig. 6.26. The former sample test with an AQL level of a certain percentage of defectives was replaced by a 100% incoming inspection. Intensive co-operation with the vendors was initiated in order to improve the quality of the components. Due to the steeply falling defect rate, it was possible to replace the 100% incoming inspection by a ship-to-stock delivery monitored by sample tests for the larger part of SSI/MSI and RAM components. It should be noted, however, that parts made in a new, more advanced technology may show much higher failure rates even today (asics up to 20 000 dpm, memories 500 dpm). These circuits, not included in Figs 6.26 and 6.27, are still 100% tested in many companies.

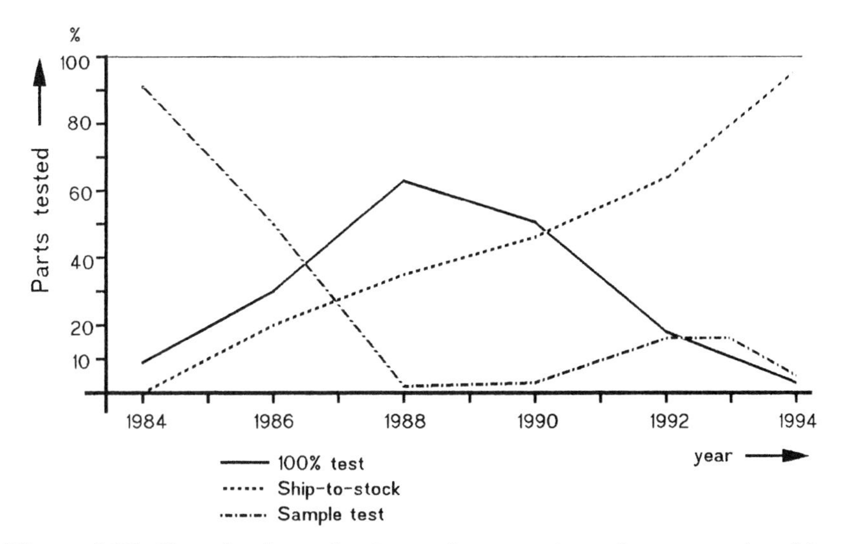

Figure 6.27 Changing incoming inspection: sample testing was replaced by a 100% incoming inspection to detect and analyse all failures in co-operation with the vendor. As soon as the failure rate diminished, testing was replaced by ship-to-stock. This is the way to achieve good quality at low cost.

6.11 FAILURE PREVENTION

The same principles for component failures are also useful for failures caused by manufacturing, such as bad soldering. This is shown by the lower cycle of Fig. 6.25.

Feedback from failure analysis to process control leads to preventive measures to avoid or reduce the occurrence of process-related failures in the future.

An analysis of the production flow shows the weak points in the case where overstress, which occurred during manufacturing, was the reason for a defective component.

A link to the system design has to allow for a design change in the case of application-related failures. The product has to be changed to take the parts specification into consideration. But this is not enough.

It is most important to improve the design rules accordingly, to prevent similar failures from being designed into new applications.

All these measures, which are explained in detail in the next sections, can be performed under a quality action as mentioned in section 6.2. Good failure statistics to recognize how far the above mentioned improvements have reduced the defect rates are also necessary.

6.11.1 Statistical process control (SPC)

SPC is one of the main methods of improving manufacturing quality. This is used with great success by semiconductor manufacturers and helps them reach excellent quality. It is well explained in the literature and in manufacturers' reliability handbooks.

The following example of manufacturing a double side mounted board shows how this method can be used successfully by manufacturers of electronic systems too. This example corresponds to box 4 of Fig. 6.4, designated by 'assembly of boards', in the total production flow.

The bare boards (copper plated, etched and pretinned) are fed into the assembly line (Fig. 6.28). The first production step is screen printing the solder paste. The next is automated placement and gluing of the components. Then the board is soldered either by infra-red heating or by vapour-phase soldering. Infra-red is preferable because of its better environmental protection. Wave soldering, the standard solder method of the past, can also be used for soldering the SMD devices of today. After the first side is soldered, a visual inspection to check correct placement and good soldering is performed.

In most assembly lines today, this inspection is done manually. Optical- or laser-based automatic solder inspection facilities will be used in the future. The image processing software necessary to identify bad solder points is not yet able to handle densely and multifariously assembled boards, according to manufacturing people, and is only satisfactory for uniform boards such as RAM modules.

X-ray transmission radiography will become an interesting alternative, especially when plastic PGA packages with unobservable solder points under their body are used. A point source emits a cone of rays which pass through the board and create a shadow image on a screen. An image at least ten times more detailed than that for lead inspection is required to interpret the results and detect all insufficiently soldered pads. The finite point size of the source limits

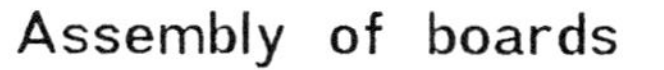

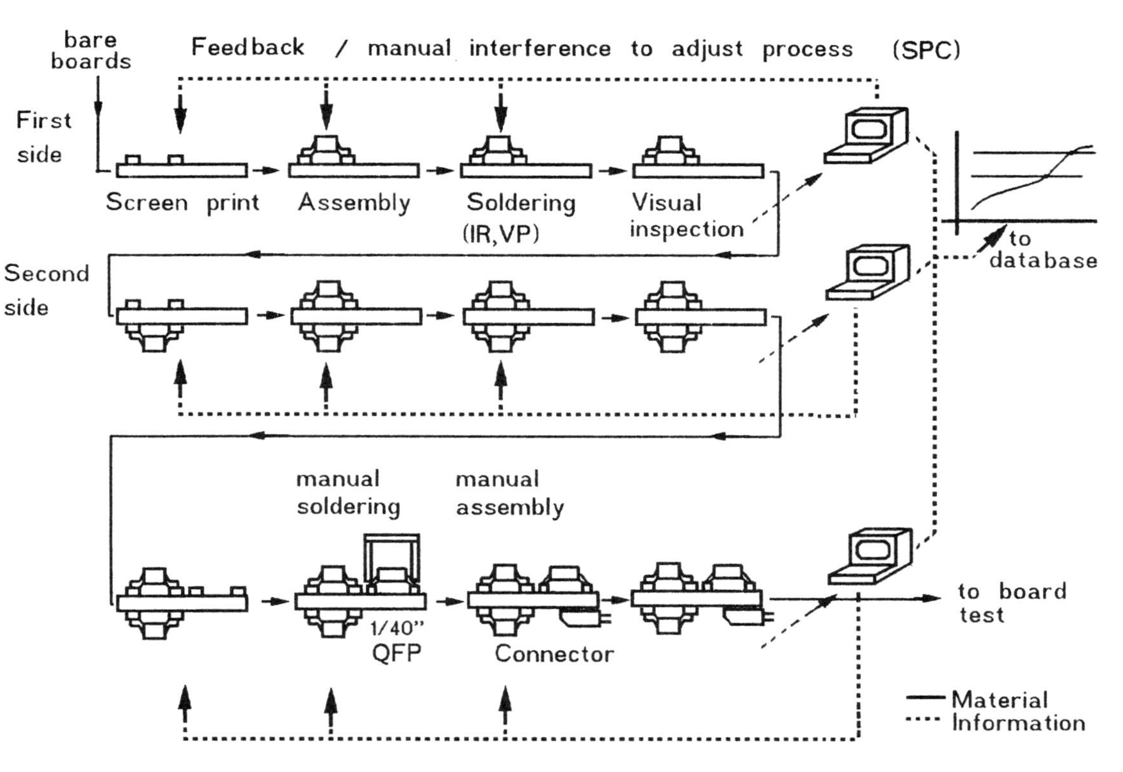

Figure 6.28 Assembly process of printed boards. A test after each production step is performed to screen out and repair defects. But statistical process control is the key to reducing the failure rate.

the image resolution. The software problems are similar to those explained above.

All detected failures are corrected by resoldering and then retested. This is not shown in Fig. 6.28 because it belongs to the manufacturing process (Fig. 6.2). The failure data (open/short, misplacement, location on the board) are fed into a database. If the number of failures rises beyond a certain limit, then the manufacturing step concerned is adjusted. This is the SPC method which corresponds to the process control of Fig. 6.2. A profound knowledge of the process and its parameters is necessary.

The same procedure is done with the other side of the board. The last step is the manual placement and soldering of advanced packages and mechanical parts such as connectors. After a final inspection the board is transferred to the board test area (box 6 in Fig. 6.4).

The manufacturing of a printed board is another example of SPC. The bare board is copper plated, etched and tin plated to prevent oxidation. After this procedure, the board is distorted by the heating process and the tin thickness is no longer uniform. Differences in height between pads on the seating plane of the same package will be up to 50 µm. The coplanarity of the package, which may be more than 100 µm will add to this. This gap has to be levelled out by the solder paste to obtain a uniform solder quality for all leads on the package. The thickness of the solder paste on the other hand is limited by the line spacing which may be less than 500 µm on dense boards. Too much solder (more than 150 µm) raises the danger of shorts between lines. The goal of SPC is to find and maintain the optimal process parameter, in this case, the solder height. This is shown in Fig. 6.29. An increase in shorts reduces the solder and an increase in opens means that more solder will be applied. The optimum depends on line geometry and package dimension, and varies from board to board. SPC regulates this.

SPC has its limits too. To reduce the failure rate further or to handle still smaller geometries, fundamental changes of the process which go beyond SPC will be necessary. In the above case, for instance, the tin plating with hot-air levelling can be replaced by a galvanized Ni/Au plating (2 µm Ni/0.5 µm Au). This leads to a more uniform surface and avoids any distortion. Such a fundamental process change however, has to be qualified and released before it is introduced into manufacturing. This will be explained later, in section 6.12.

Another procedure for reducing the soldering failure rate further is to screen out components with bad coplanarity with a 100% automated

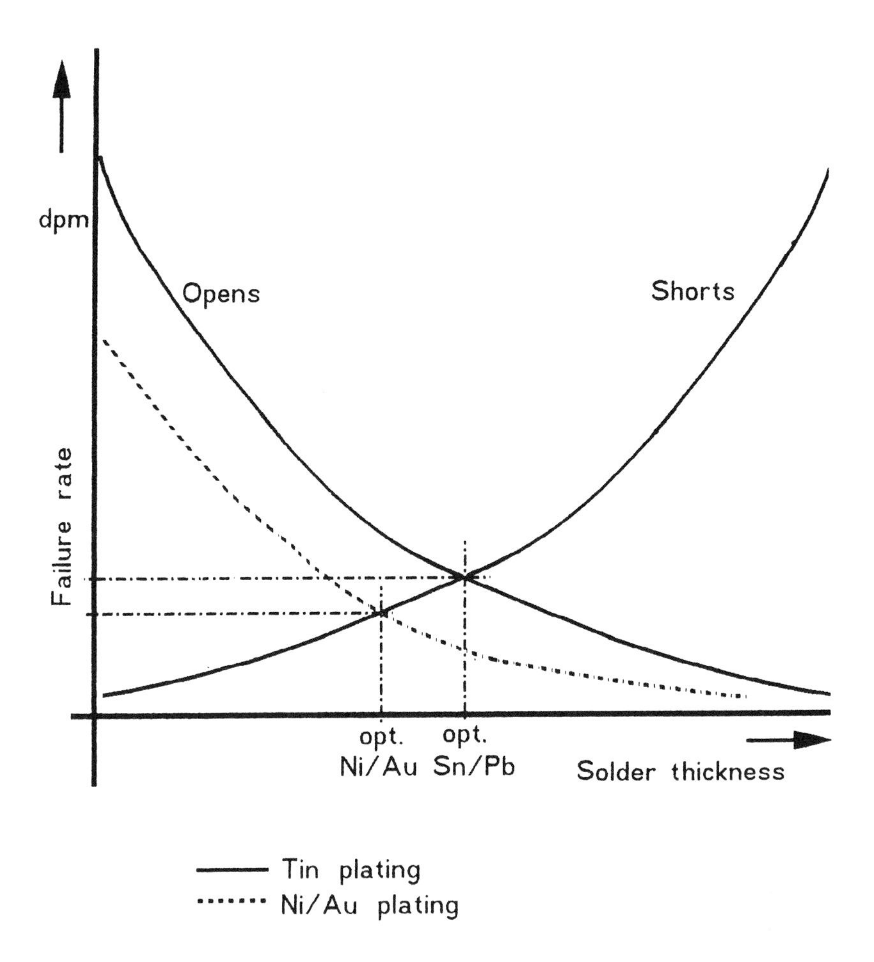

Figure 6.29 Statistical process control regulates the process towards optimal parameters.

lead inspection. Coplanarity arises from all kinds of handling: trim-and-form, testing, and placement. Lead inspection would best be performed in combination with automatic placement in order to cover all these sources, but this is as yet a promise of the future. Throughput and accuracy are important properties of the equipment. Basically, two different systems are on the market: imaging-based systems and laser-based systems. Whereas the first one captures the complete image at once and projects it on a CCD array, the new laser-based system scans each lead in sequence. Each of these system has its advantages and disadvantages. The bottleneck for both, more so for the laser-based system, is the software for image processing.

The open question behind all these plans is whether the trend towards increased fragility and miniaturization of new packages will continue. New more robust packaging techniques of plastic PGA (Fig. 5.11(b)) may eventually be of more interest.

This example of manufacturing a printed board shows the reader the operation and limits of SPC.

6.11.2 Preconditioning (run-in)

What is burn-in for the manufacturer of components, is run-in for the manufacturer of electronic systems (Accumolli, 1993).

Bad soldering is not only one of the most frequent manufacturing faults but also often a latent or dormant flaw in the boards which can remain undetected by board test and become patent later, sometimes as an intermittent failure. Monitored screening is effective in screening out these defects. Either vibration or thermic cycling can be used to precipitate these latent defects.

For vibration screening, an impact excitation has to be used which excites at all modes and at time-varying amplitudes. All resonant frequencies of the board have to be excited, and this can be performed using pseudo-random sine waves. A conventional triple axis electrodynamic shaker vibrator is not sufficient.

For thermal cycling, high thermal cycle rates are essential. A thermal ramp rate from −40°C to +70°C within three minutes (35°C/min) is recommended. Though possibly cheaper to run, more cycles at a lower ramp rate are far less effective.

If the boards are only tested after this run-in procedure, many defects remain undetected. It is essential to monitor during all the cycling or vibration. This is one more argument for implementing self test on-board.

These tests are for boards only. Thermal tests of complete cabinets are expensive and vibration tests may damage connector contacts. To detect bad contacts between board connectors and the back plane, a kind of knocking on all boards is performed instead.

6.11.3 ESD prevention

The required extent of ESD preventive measures is an individual compromise between:

- the sensitivity of the used components (section 5.3);
- the requirements on quality (the allowable ESD-caused failure rate); and
- the cost of ESD preventive measures.

For a company producing high-grade electronic devices consisting of more than 1000 high-speed ICs at high quality (less than 10 dpm ESD failure rate at board test) the following concept for a closed system of ESD preventive measures is given. It has to be realized unconditionally and without any gap to be effective. The main items of this concept are given below:

1. **Information and training of all personnel**
   The greatest danger to electronic components are the persons handling them. They can charge up to voltages of several thousand volts (Table 6.9) and can store energy of several mWs. This energy is sufficient to destroy nearly all modern electronic components, TTL as well as CMOS. Therefore all ESD preventive measures are useless if the insight and the motivation of manufacturing personnel is missing. It is most important that these people are strongly informed about the danger of electrostatic charges. Practical demonstrations and/or films are necessary to convince them. Slogans like 'static kills chips' should be posted and an easy-to-read manual distributed to the personnel. This action is not a one-off. As time passes people become careless, and this is the main residual cause of ESD failures. Constant ESD audits are necessary to keep the failure rate low (Romanchik, 1992b; Tannouri and McCaffrey, 1993).

Table 6.9 Typical values for charge-up voltages of persons

Action	*Average charge-up voltage at relative air humidity (V)*	
	10–20%	*65–90%*
Walking across a carpet	35000	1500
Working at a bench	6000	100
Handling plastic materials	20000	500
Moving on a chair with polyurethane foam mat	18000	1500

2. **Protected working areas**

Processing of electrostatic-sensitive components and devices has to be done in protected working areas only. These areas are strictly off-limits for all persons without conducting clothes and shoes. This is valid for upper management too. Protected working areas are characterized by:

- a floor with an electrostatically dissipative surface and a resistivity between $10^6\ \Omega$/square and $10^{11}\ \Omega$/square;
- workbenches and tables with properly earthed electrostatically dissipative surfaces and attachments to connect wrist-wraps;
- chairs with conductive seats which are in contact with the floor by means of metal rolls on their legs;
- earthed lockers with non-chargeable insertions;
- wrist-wraps or bracelets earthed through a built-in resistance of $10^6\ \Omega$/square;
- tools, instruments and the tips of solder irons which are earthed, soldering irons with a zero point switch, and static-safe solder suckers;
- all persons inside this protected area wear conducting clothes and shoes which have visible colour markings.

Rotating parts or tapes are a main source of danger. Keep a watchful eye on conveyor belts in the assembly line.
Missing wrist-wraps, unsafe benchtops and non-observance of regulations are responsible for more than 75% of ESD hazards.
The use of air ionizors is still unusual in many countries. This low acceptance may come from their low effectiveness and rather short operational range, and from the ergonometric and security requirements necessary.

3. **Handling and transport**

- To avoid CMD failure mode, components should not be directly earthed. This includes any contact with metallic objects with capacitance comparable to or greater than the component capacitance.
- Components must not be exposed to dangerous electrical or magnetic fields (e.g. monitors) to avoid field-induced failures.
- Packaging materials in direct contact with the component pins have to be permanent volume conductive with a specified surface resistance.

- For transport through unprotected areas, packaging materials have to be used which can screen electrostatic fields.

4. **ESD audits**
Internal ESD audits have to be performed periodically to control the effectiveness of the preventive measures and to locate trouble spots. An ESD supervisor and/or an auditor has to be appointed. An auditor cannot control the obedience of the manufacturing personnel in observing the preventive measures, but has to ascertain that the actual production facilities give maximum protection against electrostatic damage. ESD control equipment, like other equipment, needs regular testing and maintenance. The auditor has to ensure that:

   - all areas where sensitive components are handled are designated by a sign or a warning notice at all entrances;
   - dissipative floors are provided throughout, workbenches have dissipative laminated tabletops, stools and chairs are all earthed even after alterations of the buildings;
   - earthed wrist-wraps are available at each workplace and are in good condition;
   - solder irons, suckers and other tools are earthed;
   - volume-conductive boxes and storage bags are available; and
   - conducting smocks are available for visitors.

Last, but not least, are the auditors who goes around and it is their duty to convince and train people of the importance of ESD control. After each supervision, the auditor writes a status report to the management (Romanchik, 1992b).

6.11.4 Use of human resources

It is obvious that all production supervisors and workers should have thorough training in the methods and skills necessary to perform their tasks, i.e. operation of instruments, tools and machinery, reading and understanding the documents, and awareness of the relationship of their duties to safety in the workplace.

In addition, they need to be motivated towards quality and be aware of the necessity of proper job performance at all levels. The effects of poor job performance on other employees, customer satisfaction, operating costs and the economic well-being of the company should be kept in mind. It is the customer alone who decides where and what to buy and the employees' salaries depend on this decision.

It is often proposed that to encourage people to produce good quality, management should provide recognition of performance, in other words, in awards or in money. The author has some doubt as to the long-term success of such a process. Competition, instead of co-operation, is the consequence. Workers will try to hide failures. Social competence, which is an urgent need in modern manufacturing, is not encouraged.

A better solution is to install a teamwork procedure for improving all parts of the manufacturing process. Each part, starting from scheduled purchasing of components and ending at shipping of the final product, is treated as relating to internal or external suppliers and customers. The team, i.e. managers and employees, has to detect controllable parts of the process in their section of the production flow. The team has to:

- give a structural description of each process;
- estimate its influence on the quality of the final product;
- determine the goal of the process;
- determine the decisive factors for its success; and
- define checkpoints to measure the efficiency of the process.

This leads to a continuous improvement of the process. In regular team sessions, proposals are discussed and, if consent is reached, carried out. In consecutive session, the results are to be appraised by the team. By co-operation between teams, the interfaces between successive processes converge.

This method differs from a suggestion system, or even quality circles. The goal is not to make proposals for what the management or other people should do to improve quality. The team is to become active in its own sphere of operations and be willing to act jointly to overcome barriers.

In this way, there emerge small groups of teams working towards a common goal without any rivalry. This is how to use the inherent attention, willingness, skill and commitment to their work to improve quality and to reduce failure rate (Ishikawa, 1985).

Many Western companies look at the success of Japanese total quality control and many reports are published on TQM and on all its synonyms and derivatives. These methods, quality management and its undoubted contribution to the tremendous improvement of quality in the past decade are not covered by this book. The reader can easily find discussions in recent publications. However, this book will encourage the reader to ponder on the relationship between the success of a specific way of organizing industrial processes and the cultural background. Merely copying the Japanese methods may not be the best way to become competitive. The return to the cultural roots of personal engagement, towards an idea without focusing on immediate personal profit, may be a more rewarding perspective.

The future of a company may become dependent on the intuition and on the origination of ideas, which will lead to innovations of the production process. The winners of tomorrow might be companies not relying solely on a network of computers but also on the ability of their employees to communicate. If managers encourage the personal deployment of their staff, then their staff will recognize open (i.e. unregulated) requests for action and react creatively in such situations. Many failures caused by an emotional unconcern will thus be prevented. Activities of the management to quantify this effect and to use the human resources of their staff systematically are still atypical. This includes the selection of junior staff not only by his or her success charts. Ethical values may gain more consideration in the future. The quality of an employee's work depends to a great extent on his personal conditions, such as existential security and absence of mobbing.

There is no doubt that the organization of a company has a direct influence on the quality of its products too. So, a strong requirement exists to optimize management. At present, a self-organizing network is thought to be a better solution than the hierarchical structure of the past. It might better respond to the rapidly changing demands of the market and encourage creativity. It seems to be too early at the moment, however, to give a well-founded prediction of the long-term success of this re-engineering process.

6.11.5 Permanent control and calibration of test equipment

All test equipment, incoming inspection testers (ATE), production testers, board testers (ICT) and all other test systems used in the manufacturing area have to be controlled and calibrated periodically. To control, in this sense, means to check all properties which are relevant for the correct function of the test equipment. If the test results deviate from the measurement standards, then the equipment has failed and a calibration procedure has to be performed.

The scope of the test equipment of a company is often so large that some kind of software assistance is necessary. A database has to be used to store all relevant data and to organize control and calibration of all test equipment.

The reasons for the control procedure described below are:

- to assure a constant quality level of products;
- to fulfil legal requirements (e.g. product liability law or the bureau of standards) – consistent documentation of calibrations helps to exonerate the manufacturer;
- to fulfil contracts with customers in respect to ship-to-stock or other reliability programs;
- to assure against warranty claims; and
- to be in line with standard regulations such as ISO 9000.

Besides these legal or contractual reasons, there are economic or organizational requirements which demand the control of test equipment and the use of a database:

- optimization of calibration intervals;
- company-wide registration of test equipment, which gives the management an overview of available testers in different departments, their usage and status, and helps to avoid double investment;
- the provision of perfect equipment which avoids uncertainty caused by an undefined calibration status which makes results doubtful; and
- the replacement of unserviceable equipment on time.

The goal is to ensure that all test equipment used in manufacturing and testing of products is registered, permanently controlled, regularly calibrated, justified, repaired or, in the case of intolerable deviation, scrapped.

Figure 6.30 shows how to perform this recalibration of test equipment in conjunction with a database. First, as soon as any equipment is procured, the registration data (e.g. the serial number of the device, the date of procurement, etc.) have to be entered into the database. Next, the equipment has to be qualified. All relevant properties have to be tested thoroughly and a first calibration performed. The result of the qualification tests and the calibration also have to be stored in the database. Then, one has to set up a control and recalibration plan, to fix the time interval between two successive calibrations. This plan and the results of all subsequent calibrations

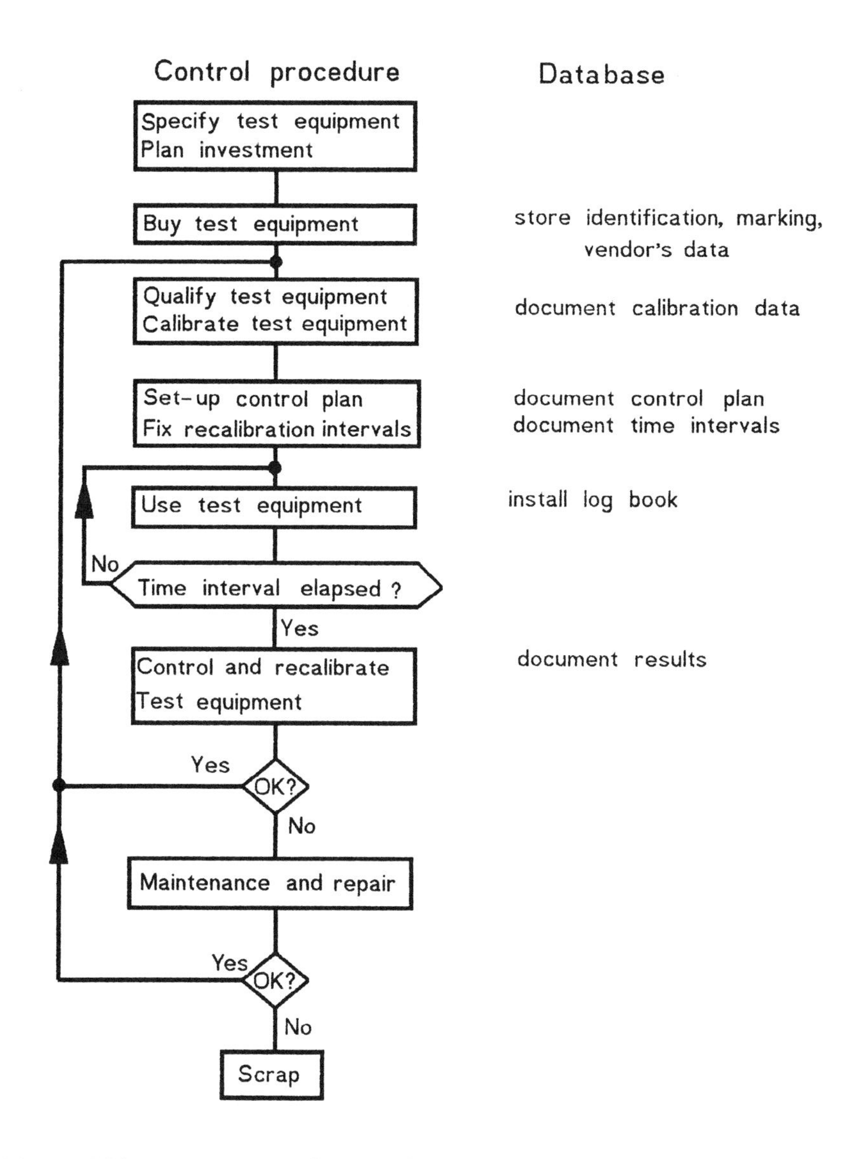

Figure 6.30 Management of test equipment.

and controls are stored in the database too, as well as all maintenance and repair work. So an optimal calibration interval can be derived from the measured deviations or can be changed dynamically. If the elapsed time exceeds this fixed interval, then an alarm warns the test personnel. This can be performed by a special hardware addition or built-in software.

During the use of any larger test equipment like an ATE system, a log book of all extraordinary events has to be kept. These may be dubious measurement results or strange malfunctions. This log also contains the date of each recalibration and is always beside the tester. It informs the changing test personnel of the latest peculiarities and also helps to track sporadic failures.

ATE systems include built-in diagnostic and maintenance software which verifies that the tester is operating properly and pinpoints problems when they occur. Calibration routines make it easy to always keep the tester in perfect status. For other test equipment, all this has to be performed manually. In each case, written instructions on how to control and calibrate have to be provided at each tester.

From this database, statistics can be derived on the quality of the test equipment, the MTBF, the number and cost of maintenance or repair calls, the on-time, the usage in percent for different users and so on.

6.12 PROCESS CHANGES

6.12.1 Procedures for introducing process changes

One should expect that, by employing quality management as described in sections 6.2–6.5, end-product quality will improve continuously and the failure rate will fall to nearly zero. One by one, the sources of failures are detected and eliminated. First, all main failure modes are eliminated. This is most effective and improves quality considerably. Later, if it seems necessary, all minor sources are treated in the same way, which improves quality still more. But, in reality, this picture, as fine as it looks, is not true. There is a counter effect too. The constraint, which is set up by competition, forces the manufacturer constantly to develop new designs with functional properties superior to the existing ones, but also with new risks. By this, new and unapproved manufacturing methods and more advanced but new components are introduced into the manufacturing process. This causes new failure modes and consequently new quality problems. If a manufacturer, for example, has learned to produce printed boards of high quality using insertion-type components, then a break in quality can be expected when this manufacturer changes to surface mounted components. The same may happen again when changing to caseless packaging. So, a certain mean quality level will never be exceeded, because the number of problems solved by the activity of quality engineers and the number of newly arising problems balance, as shown in Fig. 6.31. Conventional quality management reacts only on failures which have already happened. To gain further improvement in quality, it is necessary to avoid failures before they happen.

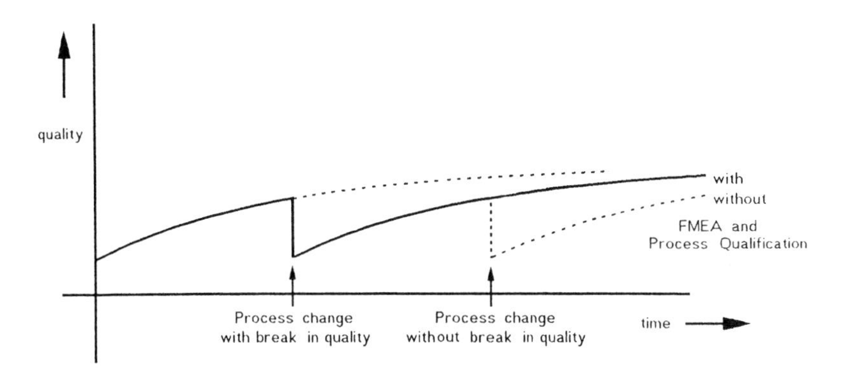

Figure 6.31 FMEA – a sudden change of the production process will often cause a break in quality. FMEA is a method of preventing this.

This is the point where two auxiliary methods come into play, process qualification and failure mode and effect analysis (FMEA).

6.12.2 Process qualification and release

All new manufacturing processes, new products to be introduced, or products which have undergone a change, have to be subjected to a qualification and release procedure. The goal of this procedure is to introduce new or changed processes into a current manufacturing line without loss of quality. The high level of quality, which has been achieved by the methods described requires a methodical procedure to realize changes. This process qualification procedure ensures that:

- the costs of introducing a new process into a running production line are foreseeable and the risks are assessable; and
- a narrow timeframe for the introduction of a new process can be set up using a precise installation procedure.

These methods consist of a three-phase procedure with an independent control in each phase, which means, in general, that all measures to be taken under the responsibility of the manufacturing department are controlled by the QA department and vice versa. This has to be performed in a written and documented form. A computer aided database is advisable.

Phase 1: Characterization

- Specification of the new process or of the process changes in writing
- Proof of their technical feasibility
- Guarantee of their adaptability into the complete manufacturing process without compatibility problems
- Check on environmental protection
- Determination of permissible parameter variation and of check points
- Simulation of the process and its effect on the quality of the end products

Phase 2: Productive testing

- Testing of process stability and tolerance of small parameter aberrations, and determination of process parameters
- Check on reliability of the end products by stress tests
- Check on controllability of the process in emergency cases
- Check on security and environmental protection
- Check on all manufacturing and QA test procedures
- Training of personnel

Phase 3: Monitored introduction and release

- Monitor the quality of the end products after the process change has been introduced
- Release to production

6.12.3 Failure mode and effect analysis (FMEA)

FMEA is a method performed in the earliest stage of development to prevent possible failures. It is carried out when a new process is planned or developed. FMEA should be started when the first sketch or description of a new product is available. So FMEA is only applicable to new products or processes, not to existing ones. Its goal is to recognize potential conceptional or structural weak points. Much experience and imagination is required from the personnel engaged in this task. The management should grant them freedom. Typically, the team comprises of experts from different departments and with different technical backgrounds. Its members include engineers delegated from design and from different manufacturing groups. The team meets many times.

The team is assisted by quality engineers whose main job is not to make decisions, but to moderate the team. They are responsible for making available all the required documents and checklists and protocol for all team sessions. Again, a computer aided database unburdens them from routine administration work like calculations and word processing, and gives them more freedom for creative work.

FMEA is not a replacement for conventional quality management, but a complement to it. Its goal is to minimize the risk involved in the introduction of new processes. Experiments explore design or process variations for reliability predictions. In these experiments, the parameters of components processes, the environmental conditions and the accuracy of tools are varied statistically.

To start an FMEA procedure, the team has to elect a team leader who is responsible to the management for the correct performance of the team sessions. The leader has to determine the number and the date of the sessions, and invite additional experts if required. Within the sessions the following tasks have to be accomplished:

- For all newly introduced elements (processes, components) all the possible failures must be listed.
- The effects of these failures on the total system must be derived by asking the question: 'How much will quality deteriorate?'
- For each failure–effect combination, possible sources must be listed.
- For each of these failure–effect–source combinations an actual available test and prevention procedure must be listed.

To demonstrate the principle of FMEA procedure, let us consider the introduction of new SMD packages with a higher lead count on boards with a reduced line width, as mentioned in section 6.11.1.

- The possible failures may be related to assembly and solder problems and to package reliability.

- To guess the expected decrease of MTBF a meeting of experts on manufacturing and on reliability tests was arranged by the quality engineers. The failures under consideration were open solder points, shorts between adjacent printed lines, misplacement of components, and reduced reliability because of thermal shock of thinner packages.
- Sources of bad soldering are large solder gaps with respect to the available solder material, insufficient positioning accuracy with respect to the smaller pad size which may give rise to misplacement of components, and evaporating moisture during soldering which may cause cracking of the thinner packages.
- Test measures are optical inspection, run-in procedures and pretest of boards.
- Preventive measures are a change to Au/Ni plating of boards, reduced coplanarity of packages, improved automatic placement and moisture-sealed shipping of the SMD packages.

To prevent deterioration of quality, all these measures have to be set into action before the production change is made. This example refers to a component and process change which happened in the past. For future changes, a similar procedure has to be performed.

6.13 ACHIEVED COST REDUCTION

Several passages in Chapters 4 and 5 guessed at achievable cost savings. Here a résumé is given. The author is aware that this undertaking will always be dependent on the situation of a company. The total quality costs are subdivided into evaluation costs and production costs.

Figure 5.12 (p. 150) showed the increase and subsequent decrease of evaluation costs in terms of the number of engineers engaged in solving quality problems. The objective was to reach the point where problems are not solved by testing but avoided by evaluation of all components prior to manufacturing.

A definite relationship between evaluation costs and the achieved cost savings in production associated with them could not be given. Since an increase in evaluation costs will not result in an immediate decrease in failure costs, the methods used by SPC are not applicable here for finding a cost optimum. Instead, some examples of costly but avoidable failures were quoted and a slow decrease in the number of quality engineers was shown.

The main cost saving, however, is in the production area, as was explained in this chapter. It is subdivided into incoming inspection costs and manufacturing costs. Figures 6.26 and 6.27 showed the reduction of failure rate of components brought about by complete incoming inspection and failure analysis, and the reduction of test effort by a ship-to-stock procedure. The test costs derived from this reduced from a maximum of 2.5% to 0.5% of total production costs whereas the defect rate decreased to below 100 dpm.

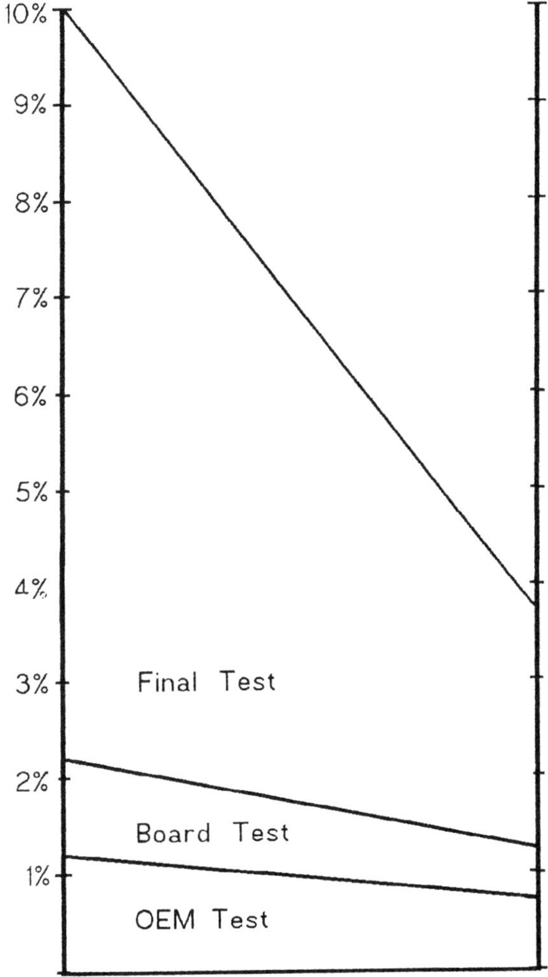

Figure 6.32 Cost of final testing. A change in test philosophy: 'detect failures at board test not at final test' and other cost-saving measures reduced quality costs by 60%.

While most companies proceeded far along the way to reducing the costs of incoming inspection, their progress in reducing manufacturing costs is less uniform. Therefore the cost reduction of board testing and final testing, shown in Fig. 6.32 from the values of Table 6.10, are derived from a few products only.

Table 6.10 Decrease in quality costs

	Start (%)	*Step 1 (%)*	*Step 2 (%)*
Test costs:			
OEM test	1.2	1.2	0.7
Board test	1.0	1.27	0.53
Final test	7.8	3.47	2.55
Total	10.0	5.94	3.78
Costs of final test:			
Rack, cable	1.56	1.56	
Design	0.78	0.78	
OEM	0.46	0.46	
Board	5.0	0.67	
Total	7.8	3.47	

All values are shown as a percentage of manufacturing costs.

The lower half shows the decrease of costs for final test.

The upper half shows the decrease of costs for all tests.

Figure 6.32 does not comprise the costs of incoming inspection and service. All numbers are relative to the total production costs and the time-scale is not graduated, because only a few manufacturers have pursued this goal to the end. Also, the spectrum of production, the economic data, inflation and other influences changed during the period of time under consideration. Formerly, there was a tendency to reduce board testing in order to save costs. The main part of quality costs had been spent in final testing. This test had to be performed manually by highly qualified people. The complex logic and the sometimes more or less sporadic failures often caused a time-consuming search procedure. The chief intention of the quality assurance management was to shift test effort from final testing to the more automatic board testing. The example in Fig. 6.15 showed the achievable cost reduction of quality costs from 10% of manufacturing costs to about 6% by shifting test efforts from the final test to the board test. As shown in Table 6.10, the cost for final testing is reduced from 7.8% to 3.47% whereas the cost for board testing rises from 1% to 1.27% in the same time. By further measures described in this chapter, the quality costs could be reduced to about 4% and a reduction to 2.5% is in view.

To perform such cost-saving measures in a production process which has to remain uninterrupted and undisturbed; a step-by-step procedure is recommended next in Chapter 7.

7

Design for quality, the key to good quality at low cost

Logic design has been subjected to fundamental changes. Electronic devices are no longer designed on a breadboard with inevitable design errors eliminated tediously on prototypes. Today, logic design is computer aided, and both highly ICs and complete electronic systems are simulated immediately. Design errors are excluded by a great deal by simulation. In reality, no designer and no simulation is perfect, so failures will still be detected on prototypes. However, the number of redesigns at the prototype stage has dropped from as high as ten to one. This design method successfully reduced time to market. The author knows of one company which finished the design of a medium-scaled mainframe one year before the expected date on using the new design method for the first time. Other companies, according to the literature, had similar experiences. Time to market was shortened from 2–4 years to 1–2 years. This did not come without a struggle. At the beginning, engineers did not believe in the feasibility of a complete simulation. Management had to insist on it until its obvious success convinced all doubters.

In the same sense, quality assurance changed completely too. It is no longer an activity in the production area for improving factory yield and reducing the failure rate of the devices in the field. It is responsible for fault-free components assembled by a fault-proof manufacturing process. Tests during production are no longer performed to eliminate bad components and to screen manufacturing failures but to check that no failure exists. The task of quality assurance in the production area is to monitor the quality of the end products.

To achieve this goal, quality assurance has to concentrate its efforts from production towards engineering. During the product conception and design phase, all key decisions which determine the quality of the end product are made (Fig. 7.1). DFM (design for manufacturability), DFT (design for testability) and DFR (design for reliability) are all terms which designate important design goals. They have to converge, however, to one goal: DFQ (design for quality). This is the essence of the thesis, that quality cannot be achieved by testing a product, it has to be designed in. Therefore the impact of all decisions made in the design phase on all downstream manufacturing phases has to be considered. This can only be achieved by a design process with institutional co-operation between design, manufacturing, test and field engineers. This was the subject of Chapter 2.

By continuous collaboration with the design group all new components, new technologies, and new test philosophies have to be compared critically to achieve optimal performance and to minimize quality risks. As the designer of a product uses simulation of logic behaviour, the quality engineer uses electrical simulation. He or she will simulate signal propagation and disturbance along lines, calculate crosstalk and ground noise and advise the logic designer on how to cope with metastability. To do this, the quality engineer has to perform a thorough examination and qualification of all new technologies and circuits, as presented in Chapters 3–5. Qualification tests include measurements of all electrical and mechanical properties, and environmental and life tests. The quality of a proposed product and its variation must be predicted and controlled by a quality simulation procedure. Quality has to be planned and should not be the unpredictable result of uncontrollable factors.

At the same time, the quality engineer has to co-operate with the manufacturing department, and has to evaluate and qualify all advanced production processes and set up a test strategy for intermediate and final tests of the planned products. A process release procedure must be launched including intensive manufacturing tests.

All this is necessary prior to a reduction of the efforts previously made in the production area. How cost savings can be achieved without loss of quality is proposed by the following step-by-step procedure. Many companies have already taken some of these steps in recent years; this book aims to encourage them to proceed further.

First step: Complete test after each production step

It is assumed that an outgoing test cannot be omitted because it has to prove the correct function of the product. Therefore, some people could conclude that a final test to clear away all failures within a complex electronic system is sufficient and the costs for the board test and incoming inspection can be saved. However, detecting a failure at outgoing test within a complex system is time-consuming, costly and requires skilled personnel. To detect the same failure at board test, or in the case of a component failure at incoming inspection, is considerably cheaper. Similar principles apply to the repair.

Therefore, a shift of test effort from incoming inspection to final test or even to maintenance will by no means reduce cost. On the contrary, faulty components have to be eliminated at the component manufacturer's outgoing inspection and not on the user's board tests. Defective boards have to be detected at board test and not at final test. So the first step should be to detect all faulty components in the earliest stage of production flow. To achieve this, tests have to be performed after each production step. Detailed recommendations on how to do this are given in section 6.4 and 6.5.

Second step: Failure analysis

The next step is the detailed failure analysis of all faulty components detected at incoming inspection or during production. No improvement of quality and no reduction of costs of quality assurance can be reached if faulty parts are only replaced in order to keep production running, although it can be tempting to do so. Real progress demands detailed diagnosis of all failures as a first step. It is absolutely necessary to document each failure with all details, with the part number, vendor and date code, the kind of tester, the test engineer and the symptoms observed, the location on board, the faulty pin and frequency, ambient temperature and supply voltage if applicable.

These documented defective parts should then be subjected to a detailed failure analysis. For this purpose, the incoming inspection tests have to be performed a second time. If the failure can be confirmed at this second test, then the defect may be caused either by some overstress during production or by infant mortality. Otherwise, if the part passes the second test, then the incoming inspection tests are incomplete or the part is not used in accordance with its specification. Because of the expense of this failure analysis, it is recommended that you concentrate at first on parts which show a high defect rate.

Third step: Correlation with the vendor

Normally, all parts are subjected to an outgoing test at the vendor's site; therefore, no fault should occur at the incoming inspection at the customer's site. The cost of incoming inspection should be avoidable.

But, in reality, this is not automatically the case. It requires much effort from the customer and vendor to reach this situation. First, trust has to be built up. Each part which did not pass incoming inspection or which failed in any stage of production has to be sent back to the vendor together with the documentation detailed in step two. The vendor and the customer have to perform a common failure analysis and they have to correlate their test procedures. The vendor should inform customers of all corrective changes in products and the customers should return the failure statistics to the vendor, so that the vendor can control the effectiveness of the corrective measures.

Fourth step: Preventive measures

The same principles are also useful for production failures such as bad soldering. Once the faulty parts have been analysed, it is easy to prevent these failures in the future. An analysis of the production flow shows the weak points, where overstress was the reason for the defect. Otherwise, the incoming inspection has to be improved, or the design of the product changed, to take the parts specification into consideration. Good failure statistics are necessary to see how far the improvements have reduced the defect rate.

Fifth step: Gradual ending of testing

The preceding steps reduced the defect rate at all production stages. Note that the cost of testing is independent of the defect rate, i.e. the cost per failure rises when the quality improves.

The rising cost per failure leads to the conclusion that considerable savings can only be achieved through a gradual reduction and, finally, ending of testing. The condition for this is the successful completion of the first four steps. The quality should be improved and a close relationship of trust should have been built up. Now one can begin to replace the 100% incoming inspection by sample tests and then, under continuous observation of the defect rate, the components should be delivered to stock without test (ship-to-stock). The next cost-saving measure is just-in-time (JIT) delivery, i.e. delivering at a precisely defined time directly to the automatic placement machine (ship-to-line). Once the placement and mounting of components and the assembly of the boards is done automatically at a proven production quality, the board and module tests can be avoided too or replaced by an integrated self check of boards. If, in the future, the intelligence of the devices increases then the end test of a device can be performed by self tests.

The objective is to deliver the components from the vendor to the customer's automatic placement machine, mounted and soldered on the boards, assembled automatically and tested by self test.

So, the investment in automatic test equipment stock and qualified test personnel can be saved; the goal, good quality at low cost, has been achieved.

Sixth step: Quality monitoring

This reduction in the costs of quality assurance is not without danger and may cause severe new problems. Small changes in the vendor's production might result in changes of some inherent properties of the components which could remain undetected by the vendor's outgoing test. The quality of the products may be adversely affected. This in turn might lead to a catastrophe at the customer's site.

Assume that the defect rate will increase from 100 dpm to 10 000 dpm (i.e. 1%) and assume that 100 parts are assembled on each board, then almost every board will have at least one defective part. The customer's production will come to a complete standstill because all test equipment has been removed and there are no skilled test personnel. The same situation may arise from an increase in manufacturing failures caused by an intentional change of the production process or by an unintentional change of production conditions.

For this reason, a continuous and fast-reacting quality monitoring system is essential. This can include testing of samples at a test laboratory. Larger companies may prefer to set up an in-house test centre, which acts like a fire brigade. Constant training is essential in this case. But these precautions have to be set up prior to the occurrence of the catastrophe. Their primary goal is not to improve the quality of components or manufacturing but to run production continuously without any problem. Even a small increase in monitored failures should cause immediate counteractions at the vendor's site either by correction of the process flow or by tightening of the outgoing tests. For this, trust and co-operation, as demanded earlier, is necessary. Furthermore, a qualified second source and a share of the supply is advisable to avoid a sudden standstill.

Seventh step: Maintaining high quality when faced with new challenges

Technological change never comes to an end. New components in new advanced technologies will be used and new, more sophisticated manufacturing methods will be introduced to maintain competitiveness in a growing world market. Permanent technological upgrading without any loss of quality is absolutely necessary to compete with low-wage countries. Many recent publications emphasize that inferior products with low costs give the lowest return on investment. Superior quality combined with advanced products gains much more return; some authors say more than five times as much.

Japanese companies gained high market share and successfully entered several markets through their quality commitment. To achieve this goal with a low-cost quality assurance system, when tests have been discontinued as a consequence of improved quality, more investment in engineering and design is inevitable, as described at the beginning of this chapter. It would be a dangerous mistake to reduce evaluation and qualification of components and processes when quality has improved. This shift of investment from manufacturing to design, as shown in Fig. 7.1, leads to a production able to react quickly to any new challenge in the market-place and maintain high quality at low cost (Brombacher *et al.*, 1993; Coppola, 1994; Dale, 1993; Elsen and Followell, 1993; Fleischhammer, 1991; Goh, 1993).

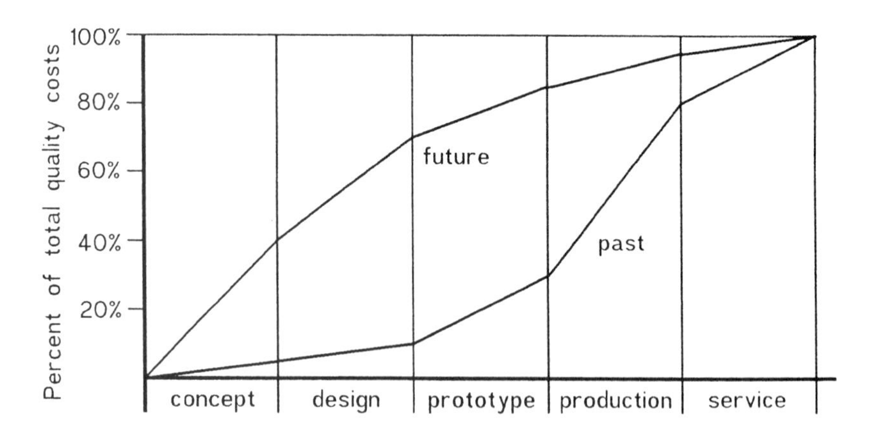

Figure 7.1 The majority of costs for quality assurance are shifted from manufacturing to design.

8

Specifications and standards as a basis for co-operation

8.1 PURPOSE AND PROCEDURE OF SPECIFYING

The preceding chapters have shown that efficient cost-saving quality assurance can only have been performed where there is strong co-operation between vendor and customer on the basis of mutual trust. This relationship has to be laid down in the form of a framework of specifications, which document the liabilities of both parties (Groh, 1993).

There are two categories of documents.

1. Individual specifications agreed upon by vendor and customer. Examples are given in sections 8.2 – 8.6.
2. National or international standards set up by organizations. Some topics of the latest standard ISO 9000 are explained in section 8.7.

The specification is not designed to obtain special selected parts, but to fix all technical and legal aspects in the interrelationships between the supplier (vendor) and customer. The most practical method is to establish a hierarchically structured order of several specifications.

8.2 GENERAL QUALITY SPECIFICATION

This contains general quality requirements to be observed on purchase and delivery of all components. It must be a constituent part of the purchase agreement or supply contract. It serves as a basis for the qualification procedure and defines the minimal requirements to be fulfilled by a vendor in order to become approved, comprising:

- the manufacturer's qualification and requalification system;
- the manufacturer's quality monitoring system;
- qualification tests and procedures;
- the manufacturer's reporting duty with respect to

- product and process changes
- problem reports
- product discontinuation;

- the customer's right to conduct a quality audit;
- the entitlements of the customer in case of a lot rejection;
- the necessary conditions for reaching a ship-to-stock agreement; and
- the general delivery requirements with respect to;

  - date of manufacture
  - marking
  - packaging
  - shipping documents.

An example of a general quality specification is given in Appendix I.

8.3 FAMILY SPECIFICATION

This defines all parameters and test procedures additional to the vendor's data book which are valid for a family of technically similar components and which are agreed upon by the vendor and customer. It contains all definitions of electrical parameters which are important for any communication between the vendor and customer. The main content should be detailed requirements of the vendor's liability to perform periodical monitoring of additional electrical parameters which are not normally tested at the vendor's outgoing test.

8.4 PART SPECIFICATION

This specifies all special requirements and/or individual parameters unique to a certain type of component and which are not explained explicitly in the datasheet. The frequency of a quartz oscillator circuit should be mentioned here as an example of an individual parameter, whereas all other parameters common to several oscillators of different frequencies are specified by the family specification.

8.5 PURCHASE CONTRACT

This contract is purely commercial and should not contain technical data. It makes references to the above specifications to define the technical requirements of the products to be purchased.

8.6 AGREEMENT ON THE QUALITY ASSURANCE (SHIP-TO-STOCK CONTRACT)

An example of a ship-to-stock contract is given in Appendix H. This contract is more a legal than a technical paper. The annexes 1–3, referred to in the example, have to be derived from vendor's information (data book, etc.).

8.7 ISO 9000 CERTIFICATION

The ISO 9000 standards have been developed by the 91 nations of the International Organization for Standardization (ISO), based on some older standards. They were first published in 1987 and have been adopted by 53 nations so far (mid-1993). These standards are universally applicable because they do not contain any detailed description. Not just computer hardware and software companies but many other industries, e.g. banks and insurance companies, comply with ISO 9000 standards.

ISO 9000 has two facets. On one hand, ISO 9004 describes standards which provide a framework for quality management and for quality systems. Although aimed at large companies, it also provides valuable input to small companies. It can help a company to develop a documented quality system.

On the other hand, the standards set by ISO 9000–9003 form the basis of a generally recognized method of quality certification. It is not ISO 9000's task to guarantee a certain quality of products manufactured by a certified company. Its goal is to certify that a company has the technical and organizational competence to use production processes and procedures which allow guaranteed quality. It is a framework for guaranteeing a company's constancy of operating a well-documented quality assurance system.

Initiated first by the European Community, it is now promulgated by many other countries too. By mid-1993, many major European and about 5% of all US companies were certified; as reported during 1994 about 50000 companies followed. There are three different certification levels available within the ISO 9000 quality certification system. ISO 9000 gives an overview and guidance through the system whereas:

- ISO 9001 covers design, development, documentation, maintenance and service;
- ISO 9002 covers production including incoming inspection; and
- ISO 9003 covers final testing.

ISO 9002 comprises ISO 9003 and both are contained in ISO 9001. Equivalent to ISO 9000 are the British BS 5750, European EN 29000 and American ANSI/ASQC Q90.

Whether you really need an ISO 9000 registration is a management decision. If a company aims to have the government or the public sector of a European country as a customer a certification may be recommendable or even necessary. NATO is requesting ISO 9000 compliance now and the Pentagon is expected to follow suit soon.

These standards force the use of a well-functioning quality system with documented quality status as a basis for continuous improvements. This is essential for all companies. Therefore, it is indeed advisable that every company obtains copies of these standards, studies them thoroughly and compares them to its own current quality system. Most of them will correspond to measures proposed in Chapter 6 of this book.

Proper control of incoming materials and components and of all manufacturing processes by inspection and test, periodical calibration of all instruments and, above all, a complete documentation of all quality-related items are central issues of ISO 9000.

Therefore, the decision for a company's management is not whether to install a quality system in conformance with ISO 9000, but whether formal certification pays its costs. ISO 9000 certification does not come free. After an initial full assessment and audit, which may last 2–5 days or more, repetitive audits have to be conducted (typically every 3 years). Surveillance audits are required in addition to this, usually annually or biannually. The costs for all this may come into the region of 10 000–30 000 US$, not including travel and living expenses for the audit team. In-house expenses for all necessary audit preparations have to be added too. This may lead to a total of 100 000–300 000 US$. On the other hand, this certification procedure gives a strong impetus to quality awareness and an orientation to the company's management and all employees. The great effort and expense necessary for this certification will emphasize the importance of quality assurance inside the company. Bear in mind that it is assumed that the importance of ISO 9000 for carrying out trade inside Europe will rise after 1996, when the EC-label becomes mandatory.

If your company decides to go for registration, choose a registrar whose registration mark will be accepted by all prospective customers. As a first step, one or more teams are installed inside the company to make all necessary preparations. This includes the permanent appointment of an internal auditor mainly responsible for these preparations and for the continuous adherence to ISO 9000 standards in between times. A self-assessment audit is a useful first step to uncovering potential weak points in the quality system which are not quite in conformance with ISO 9000. When you are fully prepared and ready for an audit, and you have chosen an appropriate registrar, you should commence informational negotiations with him or her. Send the registrar all your quality manuals and instructions for review. Be serious about any objections raised and proposals made and implement all the required changes carefully, even if they seem purely formal to you. If you are still uncertain whether your system is in conformance with the requirements, you may ask your registrar for a pre-assessment visit.

The audit itself usually starts with a meeting between the auditors and the responsible quality managers and engineers from the audited company. In this meeting, the quality system documentation will be presented and the quality managers will make it evident that they meet all requirements. Afterwards, during a visit to the production area, the assessors reassure themselves of the conformity between the supporting documents and the actual procedures. They may ask employees about their quality training or their knowledge of the principles of their company's quality system. They may check the presence of the latest revision of the quality handbook on an arbitrary workbench or they may ask the workers about the quality goals of their work. They may also check the calibration documents of measuring devices.

After the audit has been conducted, the registrar makes a critical statement, in writing, identifying all deficient areas and demanding improvements. Depending on their severity, the registrar can accept a written assurance that the requested corrections have been implemented or he or she will verify this by making another visit.

After achieving a registration certificate, your company will be included in a list of registered suppliers. This increases your competitiveness in the global market of the future and may release you from individual audits by your customers. This is most important for export-oriented companies (Hnatek, 1992; Smith, 1993).

APPENDIX A

Propagation of signals and crosstalk on interconnection lines

A.1 GENERAL REMARKS

This appendix shows an example of how a thorough electrical evaluation of components can improve the quality of the end product. Small differences in the electrical characteristics, for instance when components are bought from a second source, may cause sporadic failures in the end product by an increase in noise effects on interconnection lines between logic circuits. The available safety margin has to be observed in a high-quality product.

There are many different methods of determining the propagation of signals from the output of one electronic component, the driver, to the input of another electronic component, the receiver, along an interconnection line, and of determining the noise voltage coupled by crosstalk into an adjacent line. The latter may be the case between two printed lines running parallel on a plug-in board or between two A-wires in the same wiring channel of a back plane. The line which is actively switched between both logic states is called the switching line whereas the other line on which noise is generated is called the sense line in this context (Fig. A.1). The goal of these methods is to determine the waveform of the signals when switching between logic states and to estimate whether the receiver of the sense line will respond to the noise voltage.

These methods can be classified into several categories. One category comprises purely analytical methods, i.e. methods based on formulae which calculate the propagation of signals using the characteristic impedance, propagation time and attenuation factor of the line and the reflection coefficients of the terminations. Although these methods were the first used in this context and gave a thorough understanding of the problems, they did not gain much practical importance. The reason may be that many simplifications have to be

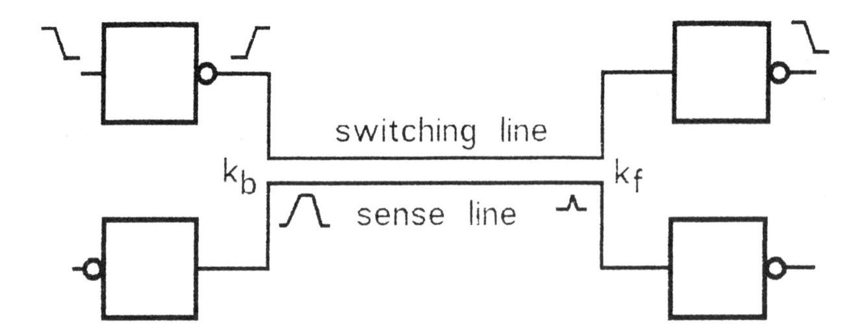

Figure A.1 Crosstalk on interconnection lines.

made in order to transfer real problems into abstract formulae so that the results differ too much from the measured reality. Above all, the nonlinearities of most electronic components are very tedious to put into formulae.

A second category is based on simulations using a model of the interconnection lines and of its driver and receiver circuits. Programs like spice are used to do this. These methods seem to be the most promising methods for the future. They are very flexible and can easily be applied to all kinds of line configurations and termination circuits. This timing simulation can be naturally integrated into the whole design procedure as an appendix to the logic simulation and automatic layout generation. The simulation programs, which can be bought from suitable software companies, may not be able to consider all the constraints imposed by technological requirements like crosstalk on the layout at the moment. However, they are rapidly improving in this respect.

A.2 THE BERGERON METHOD

A third category comprises graphical solutions. A well-known graphical method is the so-called 'Bergeron method', named after a French engineer who more than half a century ago solved nonlinear hydrodynamic problems for a public supply company by this method. This method is very simple, can be done manually without the aid of a computer and can be applied to nonlinear terminations, e.g. ICs, as well. The Bergeron method became a quick and easy way of gaining

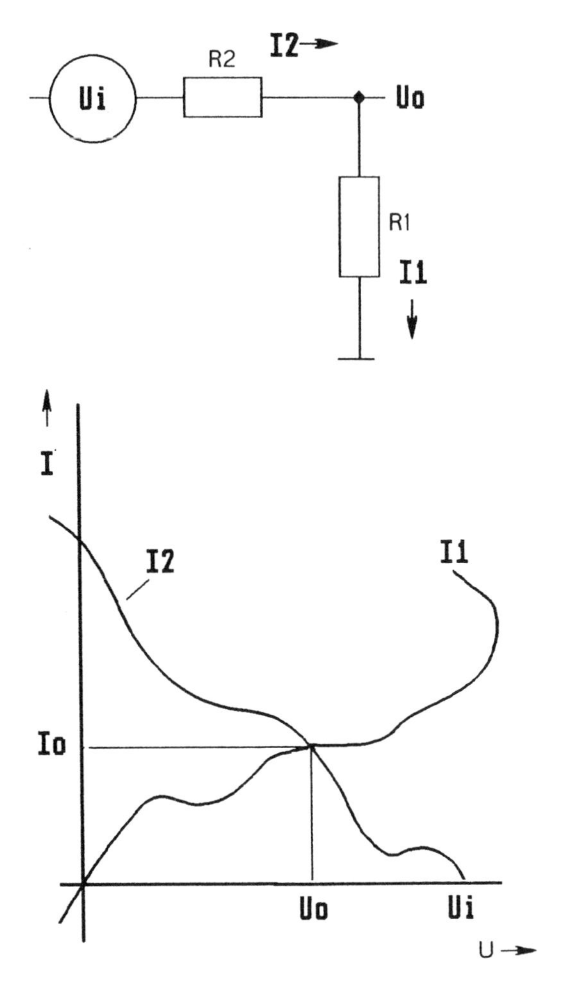

Figure A.2 Principle of the Bergeron method.

an overview of the pulse forms on interconnection lines between digital circuits and is still in use today.

In principle, this method is a graphical solution of the problem of deriving the output voltage U_o of a voltage divider which is composed of two resistors R_1 and R_2, both of which can be nonlinear. Figure A.2 shows a sketch of such a voltage divider and two characteristics:

1. The current I_1 through the resistor R_1 as function of the output voltage U_o:

$$I_1 = \frac{U_o}{R_1(U_o)}$$

2. The current I_2 through the resistor R_2 as function of the output voltage U_o:

$$I_2 = \frac{U_i - U_o}{R_2(U_o)}$$

The intersection of both characteristics points to the current I_o through the divider and to the voltage U_o at its output. This ensues from the fact that the current through both resistors is the same: $I_o = I_1 = I_2$.

Applied to the actual problem of the output voltage of a digital IC driving an interconnection line, R_2 of Fig. A.2 is replaced by the internal output resistance R_o of the driver and R_1 of Fig. A.2 by the characteristic impedance Z of the interconnection line as shown in the left side of Fig. A.3. This gives the voltage U_{a1} at the output of the driver. The same procedure is done at the receiving end of the line, where R_1 is replaced by the characteristic impedance Z of the line and R_2 by the input resistance R_i of the receiver (right side of Fig. A.3).

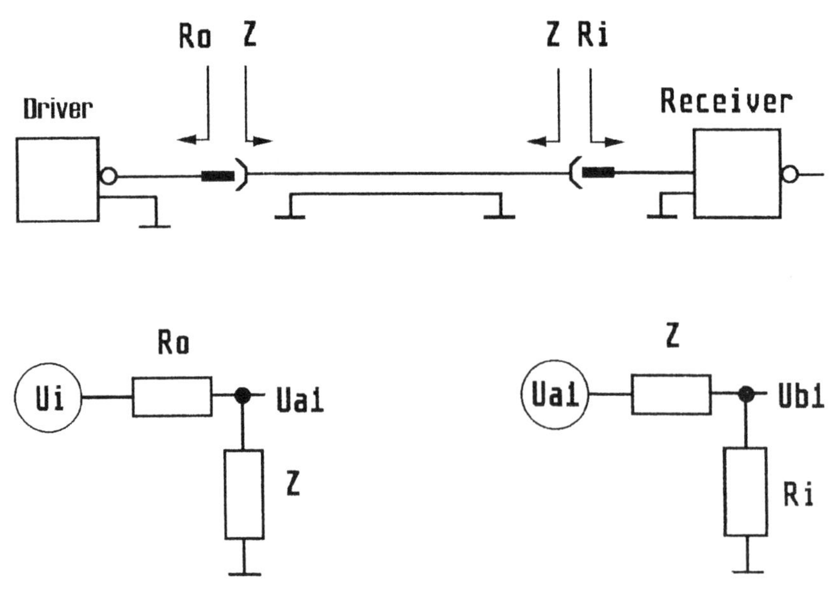

Figure A.3 The Bergeron method applied on interconnection lines.

The result is shown graphically in Fig. A.4 for switching from low to high and, in Fig. A.5, for the opposite direction. The intersection point A_1 gives the driver output voltage U_{a1} and B_1 points to the receiver input voltage U_{b1} at the first moment after switching. Now a second pulse will proceed backwards from the receiver to the driver. The same method can be applied. The intersection point A_2 gives the voltage U_{a2} at the driver output after the time $2t_d$ and B_2 the corresponding voltage at the receiver input, where t_d is the line delay. In the same way, all successive voltages at both sides of the line can be determined.

All necessary inputs for these constructions are easily available. The characteristic impedance of the transmission line can either be calculated using the geometric and dielectric properties of the line or it can be measured. The input and output characteristics have to be measured as shown previously in section 4.2. No knowledge of the internal structure of the driver and receiver circuits is necessary. Even if the simulation methods gain more and more acceptance in the future, the graphical method will maintain its significance. The design engineer has an easy-to-handle tool for estimating quickly the feasibility of a planned board design and for directly viewing the influence of changes in signal wiring on the waveform and noise.

Calculated and measured voltages and currents are compared in Figs A.4 and A.5 (parts (b) and (c)). In these figures, the benefits and limitations of the Bergeron method can be seen. This method is an easy way of calculating waveforms with a good level of accuracy (±30%) as long as the rise and fall time are shorter than the line delay.

A.3 CALCULATION OF CROSSTALK BY THE BERGERON METHOD

By the same method the crosstalk between two loosely coupled transmission lines can be determined. Loosely coupled means that the influence of the noise pulse back to the switching line can be ignored.

For this purpose, not only the characteristic impedance of the lines but also two additional line parameters have to be either calculated or measured: the forward and the backward coupling factor. A pulse running along the switching line will induce a noise pulse at the sense line (Fig. A.1). At the far end (seen from the driver) of the sense line all the coupled energy arrives at the same time. This results in a small noise spike with a width equal to the rise time of the pulse and with an amplitude which depends on the line length, the forward coupling

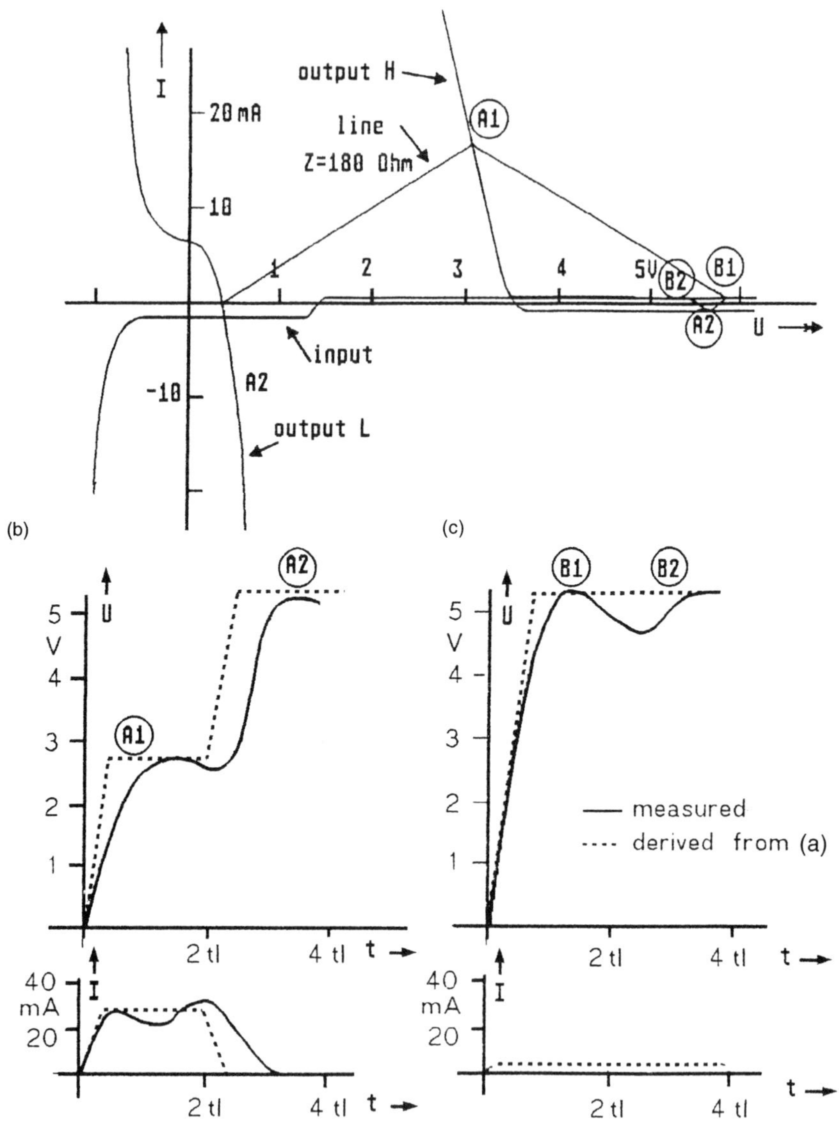

Figure A.4 Waveform of a rising edge: (a) Bergeron diagram; (b) at driver output; (c) at receiver input.

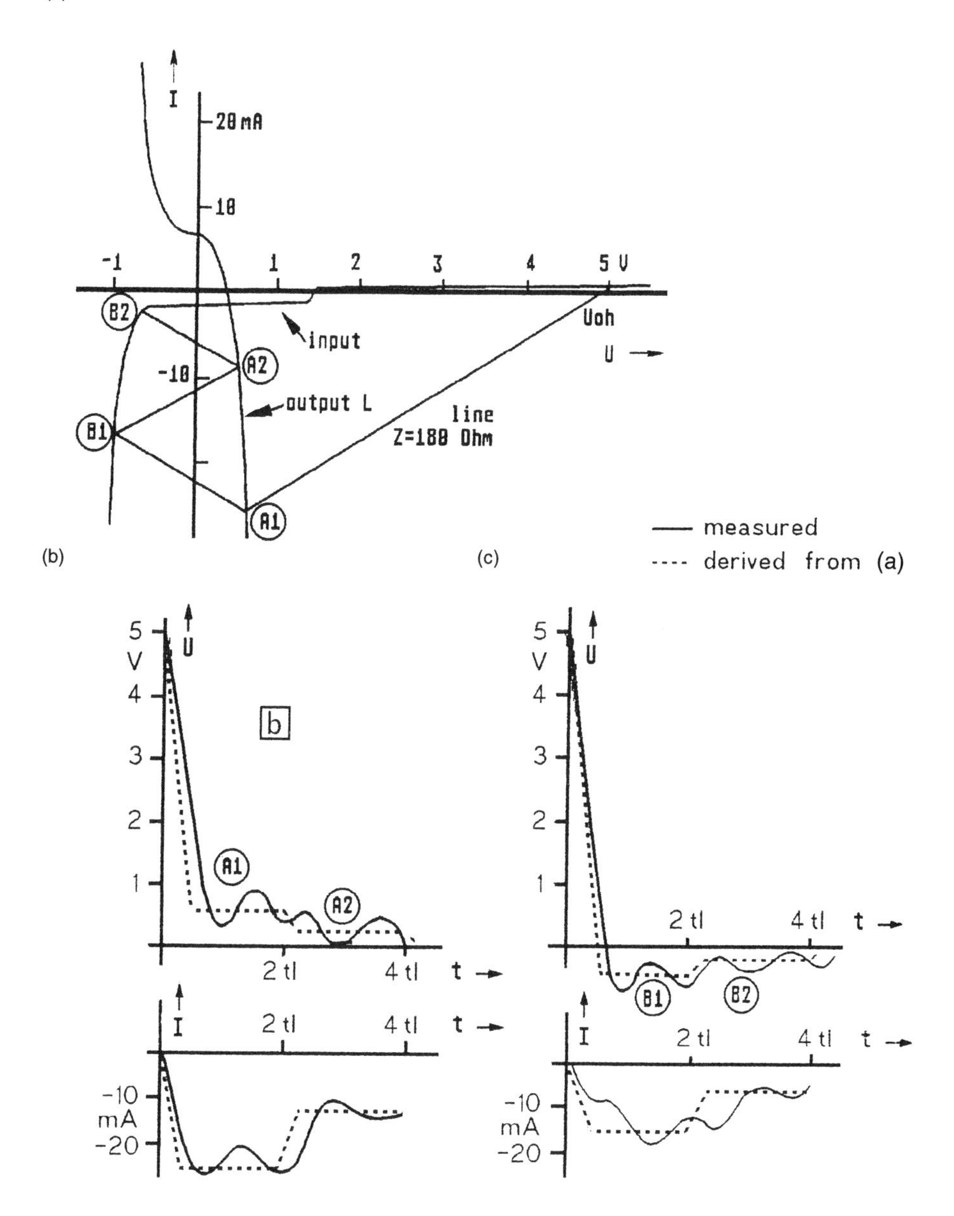

Figure A.5 Waveform of a falling edge: (a) Bergeron diagram; (b) at driver output; (c) at receiver input.

factor k_f and the amplitude of the switching signal. At the near end, the coupled energy will arrive evenly spread during the time $2t_d$. So the width depends on the line length and the amplitude depends on the backward coupling factor k_b and on the amplitude of the switching signal. To measure the coupling factors, both lines have to be terminated by their characteristic impedance, assuming that the interdependencies between the lines can be ignored due to the loose coupling. As can be seen in the following data, k_f can usually be ignored, and only k_b is responsible for crosstalk. This comes from the fact that for forward crosstalk the capacitive and the inductive components are of opposite phase and compensate. So, noise at the far end of the sense line is mostly caused by reflected backward crosstalk.

Coming to an actual example of how to derive noise on the sense line, several different cases have to be considered:

- the signal direction of the switching line and of the sense line may be in the same direction or in opposite direction, i.e. the output of the driver of the sense line may be at the near end or at the far end of the sense line;
- the sense line may be at static low level or high level; and
- the switching line may switch from high to low or vice versa.

Here only one of these cases will be shown: two wires over an earthed back plane with the drivers at the opposite side and with the sense line at a static low level (Fig. A.1). The results for printed lines would have been similar. The left half of Fig. A.6 shows the situation in the case of a rising edge on the driving line, whereas the right half shows the case where there is a falling edge. Looking at the left half (rising edge), the backward coupled noise voltage at the sensing receiver is given by point A_1. This voltage is reflected to the sending end of the sense line (point B'_2). At the same time, the backward coupled noise coming from the reflected wave on the driving line will add to this, which results in point B_1. This voltage is, in turn, reflected to the receiver as point A_2. Figure A.6(b) shows the calculated waveform at the receiving end and Fig. A.6(c) at the sending end of the sense line. The same is shown in the right half of Fig. A.6 for a falling edge at the driving line.

Graphical construction of waveforms is valid only if rise and fall time of the edges are less than two times the line delay. For a smaller line length, theoretically the trapezoidal noise pulse becomes triangular in shape. In reality, the peak of the triangle is rounded because of damping effects on high frequencies, which is shown in

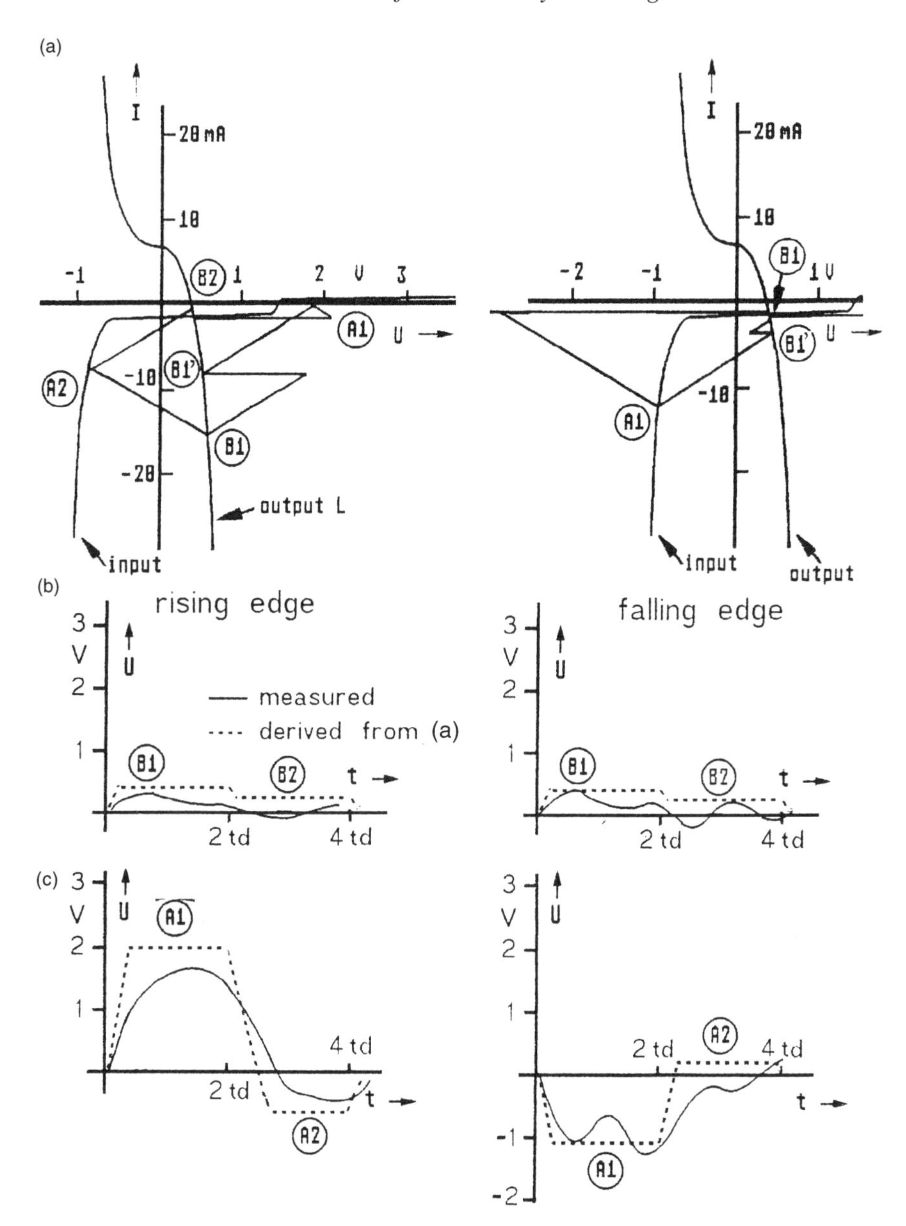

Figure A.6 Crosstalk generated noise pulse: (left side) caused by a rising edge; (right side) caused by a falling edge on the switching line.
(a) Bergeron diagram; (b) at driver output of sense line; (c) at receiver input of sense line.

Fig. A.7. This leads to a smaller effective noise voltage which means a longer tolerable coupling length or a better safety margin. Figure A.8 shows the noise pulse (height and width) derived from Fig. A.7 as a function of the coupling length compared to the tolerable noise voltage from Fig. 4.16 (page 70). This gives a realistic guess at the available margin for any specific combination of components and wiring. Furthermore, it gives an insight into the mechanism of how noise is generated and on what it depends.

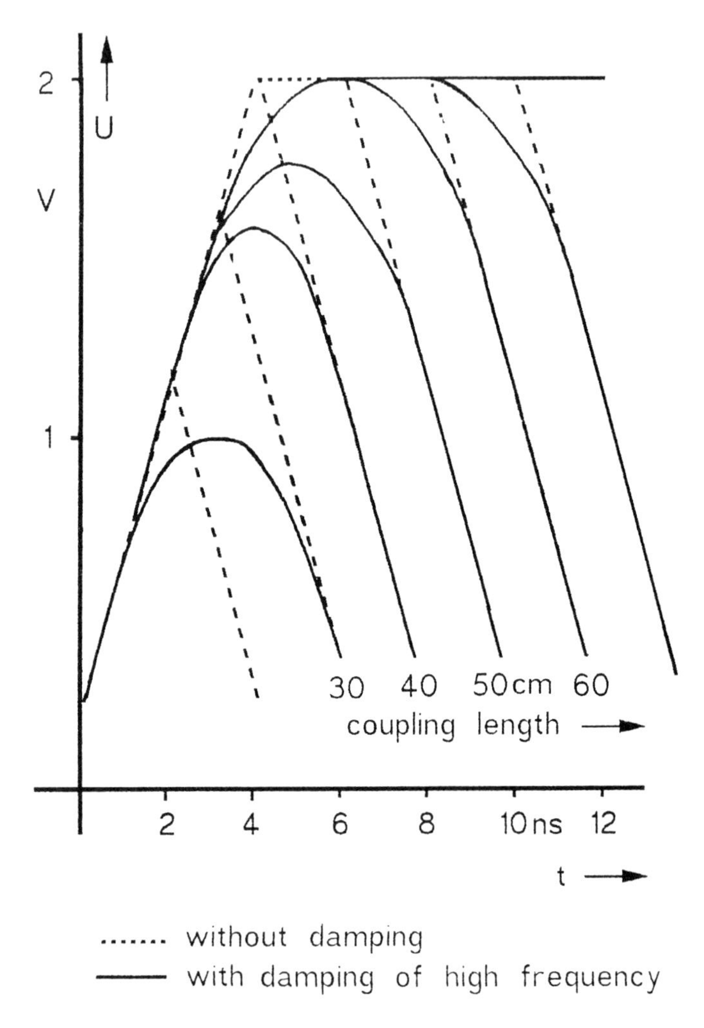

Figure A.7 Noise pulse shape as a function of coupling length.

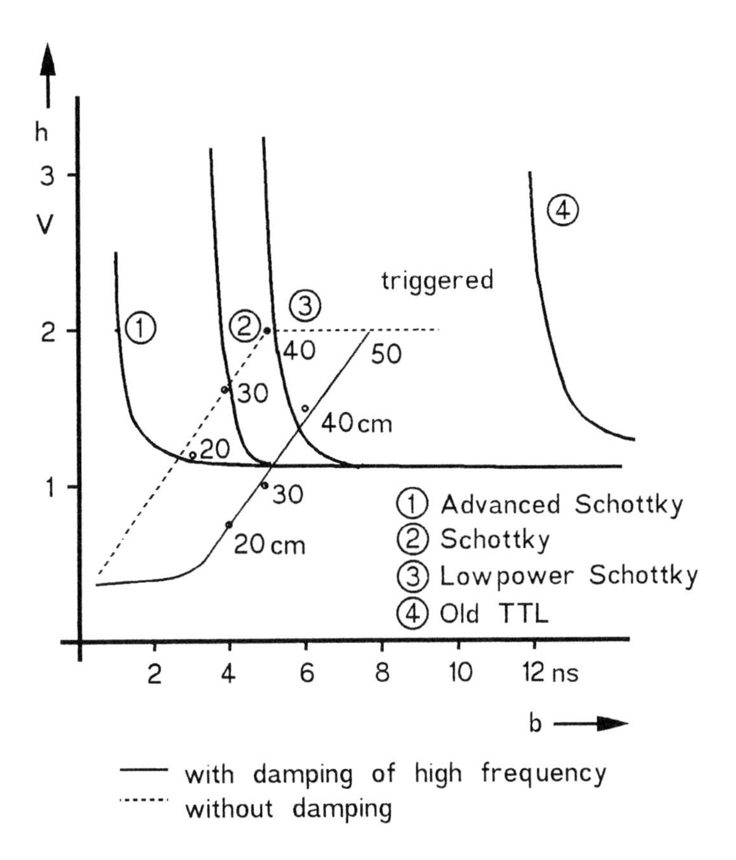

Figure A.8 Tolerable coupling length derived by Bergeron.

APPENDIX B

Evaluation test program

The efforts and costs of quality assurance can only be reduced by a shift from test and repair of the end product to an avoidance of failures from the beginning. The key to this approach is a thorough evaluation of all components before designing them in. The costs of this evaluation can only be reduced by increased automatization using software-assisted test procedures, e.g. an ATE system. One of the main problems in this context is the cost-effective generation of the necessary test programs. The aspects of test programs for evaluation are quite different from test programs for incoming inspection (Table B.1). At incoming inspection, or at the vendor's outgoing test, a high volume of parts has to be tested economically, so a short test time is a primary factor. Evaluation means a thorough test of a few samples where test time is less important. During incoming inspection, a small number of well-defined datasheet parameters are tested and the result is a go–no go statement with binning. Evaluation requires a great number of measurements to determine functions of parameters and a detailed analysis of these test results to detect possible weak points.

Table B.1

	Evaluation programme	*Incoming inspection*
Volume	Few samples	High volume
Sample	Samples of unknown behaviour outside specific points	Quality parts with specified properties
Test time	Thorough analysis with no time limit	Short test time
Number	Many measurements necessary to draw functions	Few datasheet parameters
Procedure	Variable	Fixed test procedure
Result	Detailed test report	Go–no go with binning

It turns out to be advantageous and cost saving to split the evaluation into two steps. First, all measurements are made by an automated test system and the results are stored into a file on disk or tape. Then, an analysis of the gathered data is made using a cheap computer. The software for both steps has to be generated by an evaluation program generator. This does not use a fixed test procedure. It is operated interactively by experienced engineers. For this reason, an automated test program generation of the kind described in literature is of limited use here.

For the first step, the generator should comprise a set of subroutines, each of which creates a loop of static or timing measurements. These routines are interactively called as required by the engineer and necessary parameters such as supply voltage, start and end point, pins, etc. are defined. For functional tests, the generator enables the engineer to activate a routine to learn the response of the test object to given command statements. To use simulation patterns coming from asic development would give less evidence.

For the second step, a program has to be generated which reads the stored results and compares them to the expected ones. The test engineer has to feed the specified behaviour in plain statements into the generator which in turn generates the program to perform the required calculations or other actions. If the engineer specifies, for instance, an output resistance less than 30 Ω in a certain interval, then the program draws the output characteristics from the stored measurement data, differentiates it and compares the values to the specified limit throughout the interval. The program has to decode the resulting bit pattern of the functional tests to give legible expressions of the logic behaviour. These have to be analysed manually by the evaluating engineer.

Any kind of graphical input may further simplify the program generation and help to avoid typographical errors.

To simplify the analysis of the results, the program has to allow for a hierarchical order of outputs. At first, an overview of all results is given. If certain results are dubious, then a more detailed output is required. If the overview shows for the above example a higher output resistance than specified, then the output characteristics can be printed on request. If there are still doubts, then a formatted list of the measured values is printed.

If the overview hints at insufficient test coverage, then the engineer is able to expand the test program easily by adding more command statements. An improvement of the generator itself must be possible, in case it does not allow generation of all required program steps.

An evaluation test program is not a separate tool but an integral part of the evaluation procedure and is generated at evaluation time.

APPENDIX C
Calculation of failure rate λ

Two steps have to be taken to make a rough estimate of the failure rate λ using data from the reliability test.

N devices are tested for t_t hours at temperature T_2 showing r failures.

1. Calculate the device hours (number of devices × test time) of testing and multiply it by the acceleration factor F_λ.

$$\text{Device hours } DH_t = N \times t_t \tag{C.1}$$

$$F\lambda = \frac{t_a}{t_t} = \exp\left[\left(\frac{E_a}{K_T}\right) \times \left(\frac{1}{T_1} - \frac{1}{T_2}\right)\right] \tag{C.2}$$

$$\text{Device hours } DH_a = DH_t \times F_\lambda = N \times t_t \times F_\lambda \tag{C.3}$$

where:

E_a = activation energy;
T_2 = test temperature 398 K (125 °C);
t_t = test time
K_T = Boltzmann constant 8.62×10^{-5} eV/K
T_1 = application temperature (K)
t_a = application time.

The activation energy depends on the failure mechanism. For an estimation, choose 0.8 or 0.96 eV to be on the safe side.

Table C.1: Acceleration factor F_λ for test temperature T_2 = 398 K (125 °C)

Activation energy (eV)	0.3	0.4	0.5	0.6	0.7	0.8	0.9	0.96	1.0
F_λ for T_1 = 328 K (55°C)	6.46	12	22.4	41.7	77.5	145	270	390	500

2. Calculate the failure rate λ from the number of defects and device hours DH_a. For a confidence level of 60%, use Table C.2.

$$\text{Failure rate } \lambda = \frac{\Gamma^2}{2} \times \frac{1}{DH_a} \tag{C.4}$$

Table C.2

Number of defects	0	1	2	3	4	5	6	7	8	9
$\Gamma^2/2$	0.92	2	3.1	4.2	5.2	6.2	7.2	8.3	9.4	10.4

An example of the calculation of the failure rate is given in section 5.

APPENDIX D
Reliability questionnaire

The following questionnaire is a basis for negotiations with the manufacturer or vendor of ICs for a joint evaluation. Completion by the vendor is a prerequisite for obtaining approval by the customer.

Reliability questionnaire for ICs	
Vendor Device types Packages	
Submission to:	Date:
Please return by: To customer:	

COMMENTS

To speed up the qualification procedure, this questionnaire should be completed by the device vendor. The answers will help the customer to minimize the qualification effort. In best case examples, only the characterization of technology has to be performed by the customer. For this purpose, the vendor is asked to attach five devices (representative of the series production) when returning this questionnaire.

Additional reliability tests at the customer site might be necessary if the test conditions or test results do not meet his requirements.

CONTENTS

1 General
2 Device technology
2.1 Circuit technology
2.2 Die
2.3 Assembly
3 Status of the manufacturer's reliability qualifications
4 Screening (burn-in)

GENERAL

Device type (name)
Device functionality
Package type
Marking top.......................................bottom ....

Manufacturing
Wafer line by..(subcontractor) at.. (plant/location) on...(process name/number) *A reference list of other devices manufactured at the same plant and on the same process must be attached.*
Assembly line by..(subcontractor) at.. (plant/location) *A reference list of other devices manufactured at the same plant and on the same process must be attached.*
Preconditioning (burn-in) and test by..(subcontractor) at.. (plant/location)

Process code ....
Die revision current.......................................until .... next.................................starting ....

DEVICE TECHNOLOGY

Circuit technology..

(TTL, TTLS, TTL-LS, FAST, ECL, NMOS, CMOS, BICMOS, ... ?)

Die dimensions..

minimum line spacing (μm)..

effective channel length (μm) p ch..

n ch..

die size (mm) ....

die thickness (μm) ....

Substrate material (n-Si, p-Si, n-well, p-well,twin-well, ...?)

A cross-section of the elemental devices must be attached (emitter/base/collector or source/drain)

Gate material (polysilicon, polycide, $TiSi_2$, $MoSi_2$, ...?)

Gate dielectrics	thickness (nm)	material
	....	....
	....	....

Interlayer dielectrics	thickness (nm)	material
between /	....	....
between /	....	....

Metallization	composition (%)	composition (%)
level.................. top	....	....
....	....	....
....................bottom	....	....
....	....	....

Passivation	material	thickness (nm)
top	....	....
bottom	....	....

Device technology (cont'd)

<table>
<tr><td colspan="3">Substrate material (n-Si, p-Si, n-well, p-well, twin-well, ...?)
A cross section of the elemental devices must be attached (emitter/base/collector or source/drain)</td></tr>
<tr><td colspan="3">Gate material (polysilicon, polycide, $TiSi_2$, $MoSi_2$, ...?)</td></tr>
<tr><td>Gate dielectrics</td><td>thickness (nm)</td><td>material</td></tr>
<tr><td></td><td>..............................</td><td>..............................</td></tr>
<tr><td></td><td>..............................</td><td>..............................</td></tr>
<tr><td>Interlayer dielectrics</td><td>thickness (nm)</td><td>material</td></tr>
<tr><td>between /</td><td>..............................</td><td>..............................</td></tr>
<tr><td>between /</td><td>..............................</td><td>..............................</td></tr>
<tr><td>Metallization</td><td>composition (%)</td><td>thickness (nm)</td></tr>
<tr><td>level top</td><td>..............................</td><td>..............................</td></tr>
<tr><td>bottom</td><td>..............................</td><td>..............................</td></tr>
</table>

ASSEMBLY

Lead frame
- material ....
- coefficient of thermal expansion ....

Die attachment
- material ....
- composition ....
- temperature profile ....

Bonding (ultrasonic, thermocompression, thermosonic, ...?)
- bond wire material (Au, Al, ...?) ....
- diameter of wire ....
- bond pull strength (cN) min average standard deviation

Plastic
- trade name ....
- supplier ....
- composition ....
- glass transition temperature ....
- coefficient of thermal expansion ....
- thermal resistance ....

Ceramic
- trade name ....
- supplier ....
- base material ....
- sealing ....
- sealing temperature ....

Alpha irradiation
- total flux (at die surface) ....
- sources and concentration in package material

material	^{235}U	^{238}U	^{232}Th
content (ppb)			

Lead finish
- base material ....
- composition of metallization ....
- thickness of surface metallization ....
- process of surface metallization ....

STATUS OF MANUFACTURER'S QUALIFICATIONS

The device vendor should attach the results of the qualification tests, test conditions, sample size, number of failures, target value of failure rate, failure analysis report, corrective actions on event, test location (QC or engineering)
Minimal default list of reliability tests operating life test pressure cooker temperature humidity bias for plastic packages mechanical shock, vibration, constant acceleration (for cavity packages) solderability, thermal shock, thermal cycling ESD screening (burn-in)
Additional tests HAST low temperature soft error

FAILURE MECHANISMS

For each of the following failure mechanisms information should be given on

reliability target
design measure
process control parameters
test conditions for design verification and internal qualification

Failure mechanisms

oxide breakdown (voltage)
electromigration (current density)
hot electron injection (VT shift)
ESD (voltage)
mechanical stress induced failures
metallization failures
materials mismatch (die bond)
bond crack
chip misalignment
edge coverage
package crack phenomenon

QUALITY ASSURANCE PROCEDURES

Flow of product control flow chart
Reliabilty monitoring test methods quantities per month target values

APPENDIX E

Check-list for comparing ATE testers

A check-list for comparing ATE testers is given in Table E.1.

Comments for Table E.1

1. The ability to serve two test heads gives the customer the option of increasing throughput if required. The test heads must be operable in turn, one testing, the other reloading the DUT. If different kinds of DUTs are tested, then the test time may be somewhat increased. Two test heads are practical if evaluation tests are performed in parallel with incoming inspection.
2. In the future, asics and microprocessors may have pin counts of more than 256 input/output pins. Only logic pins, not supply pins, have to be considered when defining the required tester pins. It is possible to use testers with, for example, 256 pins for parts with a higher pin count if two-pass or three-pass tests are performed (section 6.4.2, page 175). But it is advisable to buy only testers which have the option of incrementing pin count so you are able to upgrade your tester later. This is more easily achieved with tester/pin architecture.
3. Tester/pin architecture is advantageous for testing asics. The test pattern is more easily derived from logic simulation.
4. Active loads are valuable for easy loadboard design.
5. Algorithmic pattern generators are used for memory testing, now gaining importance for devices with embedded memory.
6. The ability to store serial bus test patterns (4–6 pins) without waste of pattern memory is essential for testing modern devices with LSSD bus or boundary scan.
7. Flexible pin mapping enables the testing of devices with different pinouts using the same program. Any updating of the program has to be done only once.

Table E.1

Item		*Requirements for future testers*	*Comment*
Tester			
number of test stations		2	1
number of pins/station		256–1024	2
architecture		tester per pin	3
test rate		50–500 Mhz	
accuracy		500 ps overall	
PMU/pin for parallel DC test		yes	
active load		yes	4
algorithmic pattern generator		yes	5
test bus		yes	6
pin mapping		yes	7
Test vectors			
vectors/pin		64–128k	8
bits/pin		2–3	9
subroutines		yes	10
input/output definition registers			9
masc definition registers			9
result memory/pin		yes	11
fail location memory		yes	11
match mode		yes	
automatic calibration		yes	
Available formats			
NRZ	Non return to zero	yes	
RZ	Return to zero	yes	
RTO	Return to one	yes	
SBC	Surround by complement	yes	
FIX		yes	
Timing			
minimum period		5–20 ns	12
resolution		100 ps	
accuracy			
free running clock		yes	13
external sync		yes	13
edges/ pin		32	14
placement range		2–4 periods	15
resolution		50 ps	
accuracy		100 ps	
timing sets			14

Table E.1 (cont'd)

Driver		
output high and low voltages	–2 – +8V (±15 V)	16
resolution	1 – 2 mV	
accuracy	10 mV	
resistance	50 Ω	17
programmable per pin	yes	
load current maximum	30 – 100 mA	
rise/fall time	1 ns/V (5 ns/V)	18
slew rate	1 V/ns optional	
minimum pulse width	5 ns (1 ns) at 3 V	18
input/output switching accuracy	2 – 3 ns (for tristate)	
on/off switching time	10 ns (for tristate)	
Comparator		
number of levels	2 – 4	18
compare voltage	–2 – +8V (±15 V)	
resolution	1 – 2 mV	
accuracy	10 mV	
input resistance	>10 MΩ (50 Ω opt.)	
programmable per pin	yes	
hysteresis up/down	<5 mV	
window mode	yes	19
edge mode	optional	19
minimum window size	≡ minimum pulse width	
placement range	2 – 4 periods	15
Power supply		
number of supplies	4	20
maximum voltage	8 – 32 V, 2 – 6 ranges	
resolution	2 – 8 mV	
accuracy	5 – 30 mV	
maximum current	2 – 5 A	
settling time		
maximum capacity drive capability	>50 µF	21
slew rate	~1 – 5 V/ms	21
Vbump	yes	
High-precision PMU	1 – 2/system	
Software modules		
time measurement routine		
learning mode		
splot routine		

8. Test programs, especially for asics, will be generated by program generators. These produce large amounts of test pattern because they do not usually have the ability to reduce pattern size, e.g. by subroutines. Therefore, the pattern size of the local test vector memory should be greater than 64 kbit/pin with a dynamic reload facility from a test vector store of several tens of megabytes. The data transfer from local vector memory to the pin electronics may be interleaved when testing at maximum speed.
9. Each test vector has to supply at least three bits, data, mask and input/output definition in conventional testers. Modern testers include this in the timing generation.
10. Testing of microprocessors and peripheral drivers is much improved if the local test vector memory provides the ability to program subroutines. The search for failures is simplified if the suspicious part of the test pattern is cycled.
11. Result memory and fail location memory are options which may reduce search time for failures considerably.
12. Microprocesors are clocked at more than 100 MHz now with a trend pointing towards 200 MHz. The data frequency is half the clock frequency. So a maximum frequency of 100 MHz is sufficient if the feature of doubling the frequency for one pin by ORing two pins together is provided by the tester.
13. For testing microprocessors, a free-running clock is used to test clock inputs of microprocessors, and an external sync input synchronizes the tester clock to the output of an oscillator inside the DUT.
14. Assigning timing generators to pins is often a problem for testers with shared resources. There are restrictions in some testers which aggravate automatic program generation. This is a main advantage of the tester/pin architecture.

    Two timing edges are required by the RZ format. A third timing edge is required for the SBC format and input/output switching is necessary for the test of microprocessors. But this is a minimum requirement. A more flexible architecture of timing generation which allows free assignment of more timing edges on the fly to each channel is most important.
15. Placing a comparator timing pulse within a range of two clock periods is often a requirement of a high-speed dynamic test for processors. A placement range of four periods is advisable.
16. ECL circuits operate at between –5 V and 0 V, but these levels can be shifted to –3 V and +2 V. CMOS and TTL operates between –2 V and +7 V including overshoot. Therefore, driver

and comparator voltages between –3 V and +7 V conform to both circuit types, whereas interface and analogue components often need driver and comparator levels between ±15 V.

17. Source resistance and rise/fall time are essential for ringing on the loadboard lines and for maximum signal band width. A compromise has to be made between minimum pulse width and slow edges.
18. Four trigger levels are needed to test for tristate level in one run.
19. Window mode means that the comparator input is open during a small window; this is the normal case. An edge mode comparator is triggerd by a signal edge; this should be an option.
20. Logic circuits need only one supply voltage, whereas special circuits, interface and analogue, need up to three. An additional supply for supplementary circuits on the loadboard is very useful.
21. The power supply module has to drive the necessary blocking capacitors without overshoot. It is not good practice to ask the user to program a ramping.

APPENDIX F

Principles of test software generation for ICs of high complexity

The rising complexity of ICs is responsible for the increasing attention paid to methods of testing. Despite the present discussions about the required scope of testing or the necessity of testing at all, it is generally assumed that some kind of testing is necessary to ensure high-quality economic production. It may be done by the manufacturer, by the customer or by both parties. The costs of generating the necessary test software depend on the complexity of the test object and they could constitute a considerable part of the total production costs. Therefore, methods for generating test programmes effectively are of great interest.

Several proposals for cheap test methods have been made in the past to overcome this problem:

- **Application tests** are go–no go tests. The circuit under test is inserted into an actual target system where critical application software has been launched under extreme test conditions. The problem is to define really critical applications and conditions. This kind of test might only be sufficient for a test object which is used in a single application which is never altered, such as an automotive controller circuit.
- A **golden device** is an arbitrary test program, which may consist of a randomly selected series of commands, run on a simple functional tester. The test object and a so-called 'golden device', i.e. a known good device, are tested in parallel and the test results compared immediately after each test step. The problems are how to define 'known good' and the completeness of the applied test pattern. Testing critical timing conditions and carrying out failure analysis may become difficult.
- Several papers propose a **built-in self test** of the circuits, by using part of the available gates to form special test structures or

test logic. But this will more or less reduce the performance of the circuit. The final goal of a complete self test without any external interference, will be no more than a go–no go test with little failure information. Any later extension of the test coverage means a redesign of the circuit.

The first two methods have obvious disadvantages and did not gain general acceptance in practice.

All of the above proposals have been realized in special applications only. It is still common practice to use an automatic tester to perform the outgoing test and incoming inspection of ICs.

F.1 THE TWO METHODS FOR GENERATING TEST PROGRAMS

An IC is a piece of specially prepared silicon with a pattern of metallic conductors on its surface. It would not be very practical to perform some kind of physical test, i.e. to check the doping of the silicon or the geometry of the metallic pattern. Electrical tests are performed instead. They are based on a model of the IC. It is assumed that this model reflects all properties of the circuit under test which are relevant to its correct function. The reader should keep in mind that this assumption is not a matter of course. Physical properties of the circuit which are not included in the model are not tested at all. A positive test result means that the circuit conforms to the model; it does not mean that its function is correct in every case.

The two methods for making a model of an IC, the structural method and the functional method, are discussed in the following section and their different applications are explained.

F.2 THE STRUCTURAL METHOD

The basis of this method is a structural model of the circuit, which usually means a gate model. This model is composed of ideal gates connected by ideal interconnection lines. Ideal means that these components are purely logical; no physical effects like crosstalk or other sources of noise are considered. Other primitives, like transistors or flipflops, can be used in addition to or instead of gates. Failures (opens, shorts, stuck-ats, etc.) are now inserted into this model. These failures have to be taken from a predefined failure catalogue. Test patterns, which are suitable for detecting the inserted failures, are derived from simulation, the so-called failure simulation. The advantages of this structural method are as follows:

- The test pattern can be generated algorithmically if the gate model is available, which is mostly the case for asics. This procedure can be automated and it can be performed by technicians or test engineers who need not know the logic function of the circuit in detail. The logic simulation patterns, necessary earlier for the logic design of the circuit, can be used as a starting point for the failure simulation, but this is not a condition.

  This kind of automatization has its limits too. Generating test programs by failure simulation is no push-button procedure. All software products of some complexity have bugs at first and have to be debugged, and this is the case for test software too. Debugging test programs needs detailed knowledge of the logic contents of the circuit under test and experienced engineers.
- The resulting failure coverage (the percentage of inserted failures, which are detected by the test pattern) can also be calculated automatically. This gives an objective measure of the quality of the test program. If the failure coverage seems to be insufficient, the test pattern can be expanded by additional simulation steps.

But there are disadvantages too:

- The simulated failures, at least at the moment, are purely static and the simulation is cycle oriented. A test coverage of 99% can be obtained for these kinds of failures. It is also possible to include timing failures in the failure catalogue (delay, set-up and hold time, clock skew, etc.) and to perform a timing simulation, but this requires much more effort. It seems to be difficult to achieve more than 50% failure coverage for timing failures at the moment for complex ICs. Topological failures (e.g. crosstalk between adjacent elements or lines) are included in test pattern only in very regular structures like memories.
- The number of test patterns is rising considerably with the complexity of the circuit. As a counteraction, logic restrictions are imposed on the designer in order to improve the testability. Such restrictions may be:

- do not use the inherent delay of gates for pulse generation
- no feedback on gate level
- do not use long sequential chains like counters or shifters
- provide a general reset
- integrate a scan bus
- use modular design principles.

All this will increase gate count further and cannot guarantee easy testability for gate counts > 100 000.

F.3 THE FUNCTIONAL METHOD

This method does not use a structural model (gate model) but a functional model of the circuit. The circuit is not treated as a composition of gates but as a black box (or at least as a grey box, i.e. an arrangement of several black boxes). So the basis of this model is not the internal structure of the circuit but its logical intent. Therefore, the model does not consist of an array of gates and a wiring list but of a list of commands and their parameters, to which the circuit will respond. From this list, which is given in the datasheet, a test pattern is generated. Each command is executed several times with different critical parameters and the results compared to the expected ones. Defining critical parameters requires a great deal of experience. Therefore, generating a test program on a functional basis has to be done manually by experts. But, today, logic design is also done manually at gate level and, in the future, will be more and more software assisted by a design language. To the same extent, the functional test program will be derived from this design language in the future and experts will be replaced by an expert system.

The advantages of the functional method are as follows:

- The number of test patterns necessary to obtain a certain test coverage is much less than for the structural method – this factor depends on the difference between the number of internal nodes and transistors compared to the number of functions which are defined (e.g. for a floating point unit an improvement of more than 1000 could be reached, but usually the pattern size is reduced by a factor of three to 30).
- The number of tests can be easily adapted to the requirements. If some tests never fail, then it is easy to omit them, whereas other tests can be added without difficulty if the test coverage is insufficient for some functions.

- There are no requests to improve testability with added test hardware as in the scan path method. That is because only the normal functionality of the circuit is tested.
- For the same reason, it is easy to test circuits containing mixed logic, analogue and digital, on-chip.

The disadvantage is, as explained, the lack of automatization. There are efforts underway to create the pattern interactively, computer aided using a special command language. Expert systems can be used but so far with no real known success.

F.4 RANGE OF APPLICATION OF BOTH METHODS

Both methods are used today to generate test programs; the structural method mostly for asics, the functional method for standard circuits such as microprocessors or memories. The reasons for this division are given below:

- Test programs are used by the manufacturers of ICs and by their customers. The manufacturer uses them to identify bad samples at all stages of production and for quality control. The structural method is best suited to this purpose because the patterns generated hint at the location of defects on the chip. From this, it is easy to pin-point on the weak points of the production line and thus to improve them. Most defects at this stage of production are stuck-at failures which are completely detected by structural patterns.
- At final test on the manufacturer's site, patterns generated by the functional method are the better choice. These patterns are well suited to detecting timing and parametric failures which remain after stuck-at defects have been mostly eliminated. They are also economic. The smaller number of test patterns and the adaptation of test scope to real needs reduces test duration, which becomes important for mass production. Many manufacturers, who started final testing with structural patterns derived during logic design, changed to functional pattern later on to minimize test expenditure.
- The test program is used to define the interface between manufacturer and customer. It arbitrates all dicussions concerning quality or rejection of defective parts. In this respect, there is a big difference between asics and standard circuits. If

the customer does the logic design of the asic then the manufacturer's responsibility is only to guarantee the function of all gates and the correctness of the wiring in accordance with the customer's wiring list. This is exactly what a test program on a structural basis does. Therefore asics are the principal domain of structural test programs. Because the functional behaviour of an asic is the user's responsibility alone, functional patterns would not be a good manufacturer–user interface.

For standard circuits, on the other hand, the logic design is done by the manufacturer. A datasheet specifying all functions is the user interface. The internal structure is totally the responsibility of the manufacturer. A second source, while specifying the same functionality, may have a quite different internal structure and manufacturing process. For these reasons, a test program generated on a functional basis is the better choice in the case of standard circuits. This is also valid for asics if they are designed by the manufacturer to a customer's specification.

Some authors claim that functional patterns have to be used for program generation when the internal structure is not known. But although this is correct, it is not the main reason.

- A test program is important for failure analysis too, if an electronic device fails at final test and components are suspected of causing the failure. Functional patterns are far better if the failure occurs sporadically or under certain conditions only, because the critical patterns can be easily looped.

F.5 FUTURE TRENDS

There are two methods in use to predict the future:

- **Trend analysis** The actual trend is extrapolated into the future, taking all known external influences into consideration. This method is widely used for speculation on the stock exchange. Its limitation is that sudden unexpected breaks cannot be predicted. The present trend for semiconductor circuits shows a rising number of asic designs; even standard circuits are often designed like asics by the manufacturers. The structural method of generating test programs prevails and is gaining more and more importance at present.
- **Analysis by analogy** This method compares the present situation with similar situations in the past, or in other areas, with

comparable problems in order to make predictions of the future. This method assumes that the course of evolution follows some basic rules. One of these rules is the equlibrium of antitheses. If two complementary methods are available and one of them is going to become prevalent because of some favourable circumstances, then the other one will make up rapidly as soon as the situation changes. That means the present trend towards structural pattern generation may discontinue and the functional method could become prevalent if, for example, an algorithmic program generator becomes available.

APPENDIX G

Shielding effectiveness of a cable connection

Two kinds of EMI have to be distinguished: noise generated or picked up by electromagnetic radiation, and voltage drops between different earth connections within a spatially distributed electronic system. An excellent method for measuring the radiation generated by an electronic system is based on RF transmitters and receivers (MIL-STD-285 and MIL-G-83528). The transfer or coupling impedance is a measure of the ground noise which is coupled from an external noise source into an electronic system (ARP-1173).

Figure G.1 shows two separate cabinets of an electronic system connected by a shielded cable. An external ground noise current I_{ex} generates a voltage difference U_n across the ground impedance L_g between the two cabinets. This voltage will be effective as noise voltage at the input of the receiver in the case of an unshielded cable. If the cable is shielded then this voltage U_n will cause a current I_s flowing through the cable shield. By electromagnetic coupling, a

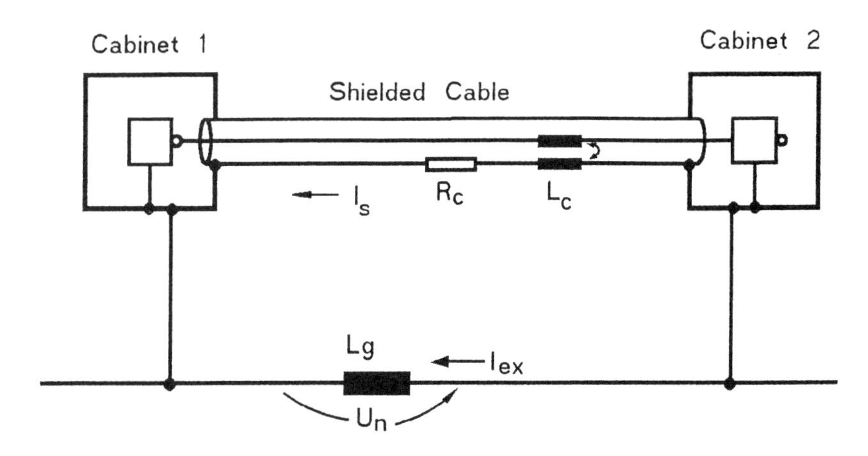

Figure G.1 Explanation of coupling resistance.

voltage of about the same magnitude is induced along the signal line inside the cable. It will compensate the noise voltage between the two cabinets if the shield is perfect. The imperfection of the shield results in a smaller voltage induced along the signal line and the compensation is not perfect. A residual noise voltage remains. This imperfection is presented by a hypothetical ohmic coupling resistance R_c. To define the total coupling resistance, the connectors at both ends of the cable have to be considered.

The coupling or transfer resistance R_c is calculated thus:

$$R_c = R_c(\text{con}_1) + R_c(\text{con}_2) + R_c(\text{cable}) \times l(\text{cable})$$

The noise reduction factor R_n is:

$$R_n = \frac{R_c}{(R_c + \Omega L_s)}$$

with $L_s \sim 1-2\,\mu\text{H/m}$. The resulting EMI-noise voltage U_i at the receiver input is:

$$U_i = U_n \times R_n = \frac{U_n \times R_c}{(R_c + \Omega L_s)}$$

To measure the resistance R_c, a piece of cable (e.g. 1 m long) is used which is shorted at one end. A current is fed into the cable shield and the voltage between the inner conductor and shield at the other end of the cable is a measure of the transfer resistance R_c. This resistance is of interest in the frequency range 0.1 – 100 MHz. Because of the high effectivity of the shield, the voltage is usually very small and difficult to measure. Capacitive coupling may falsify the result, if the test set-up is insufficient and experience is lacking. Therefore it may better to delegate this measurement to a test laboratory. Not only is the transfer resistance of the cable essential but that of the connectors as well. The transfer resistance of a connector should not be more than that of 1 m of the cable as a rule of thumb. Figure G.2 shows examples of measured transfer impedances of connectors and cables.

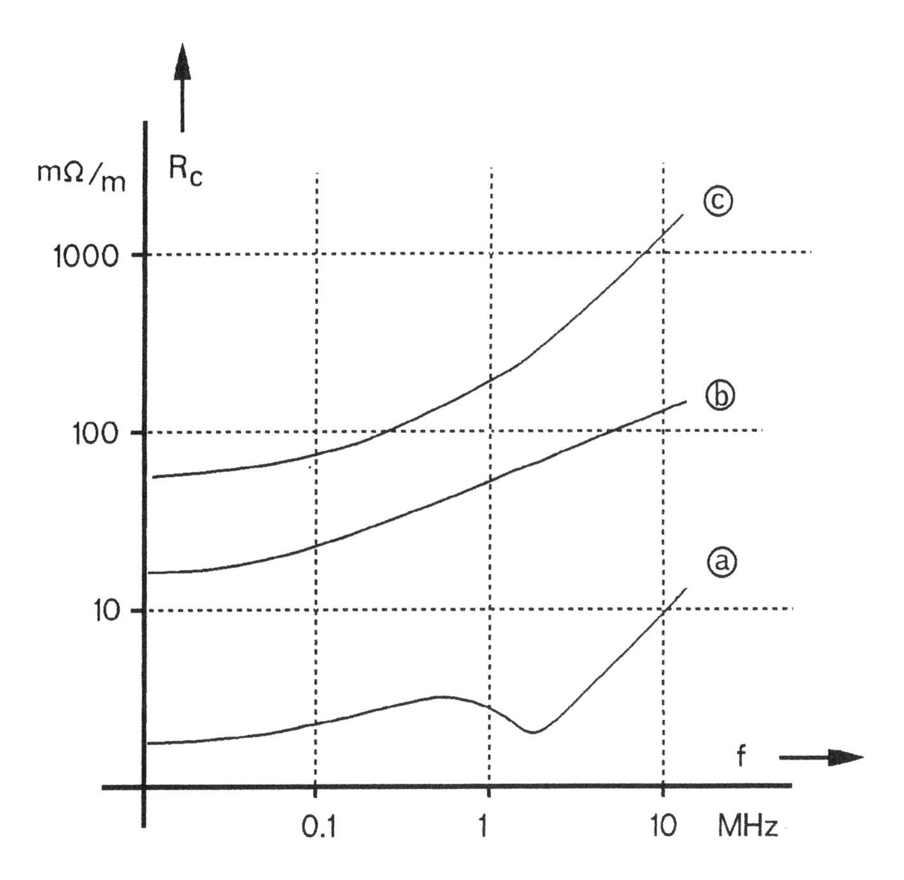

Figure G.2 Examples for measured coupling resistance of 1 m of a shielded cable: (a) good cable; (b) fair cable; and (c) marginal cable.

APPENDIX H

Agreement on quality assurance between purchaser and supplier

The purchaser knows the supplier as a quality-oriented and reliable manufacturer. The purchaser needs the supplier's products and intends to purchase them from him on the basis of separate contracts.

Both parties are agreed that the high quality and reliability of technical products in the face of undiminished competition can only be achieved if there is improved co-operation between the manufacturing stages, the quality assurance system is applied, test methods laid down, lead times shortened and test duplication avoided.

The parties are further agreed that the purchaser can cease incoming inspections to render 'ship-to-stock' or 'ship-to-line' deliveries feasible.

This being understood, the parties agree as follows:

1. **Items constituting the subject of the agreement**
   - 1.1 The items are described in Annexe 1 of the agreement. The supplier has checked the descriptions in Annexe 1 and found them to be conclusive.
   - 1.2 If the purchaser has left a sample with the supplier or if the supplier has manufactured a sample accepted by the purchaser, the design and properties of this sample are deemed to be of the kind due and owing under the agreement, in addition to the description of Annexe 1. If there is indeed a sample, the supplier will have checked it for agreement with the description and regard such an agreement to exist.
   - 1.3 If in the course of development, manufacture or testing of the items, the supplier finds the description in Annexe 1 to be incorrect, open to misunderstanding, incomplete or at variance with the sample, it will immediately notify the purchaser in writing and submit suggestions to it for remedying the situation.

2. **Implementation of quality assurance**

2.1 The supplier will maintain a quality assurance system which meets the particular requirements stipulated in Annexe 2 of this agreement. It will manufacture and test the items in conformity with the rules of this quality assurance system.

2.2 The supplier will test the items in accordance with the provisions set forth in Annexe 3 of this agreement, prior to each delivery to the purchaser. The supplier has reviewed these provisions for conclusiveness and compatibility with the tests as required by its quality assurance system, and has found this to be the case. Section 1.3 of this agreement shall apply accordingly.

2.3 The supplier will suitably mark the items or otherwise ensure that whenever an item is found to be defective, it is possible to determine the total number of items affected or possibly affected. The supplier will keep the purchaser informed of its marking system to enable the purchaser itself to identify defective items at any time.

2.4 The supplier will keep records (documentation) of the quality assurance measures it has taken, in particular for measured values and test results, and will hold them as well as any samples available in a neat and orderly manner. The nature, extent and length of period for keeping these records and samples are described in Annexe 2 of this agreement.

2.5 On first request, the supplier will grant the purchaser the right to inspect its records (2.4) in their entirety and to surrender any samples desired. It will also assist the purchaser in the evaluation of records and samples.

2.6 The supplier will grant the purchaser's representative access to his factories and premises, as may be required for checking the supplier's quality assurance system and its operation (quality audit). If the supplier purchases components for the items from subcontractors, the supplier will place these subcontractors under the contractual obligation, if so agreed with the purchaser, to accept and allow such quality audits on their premises as well. The purchaser will give advance notice of his representative's visit.

3 **Information**

3.1 The purchaser will notify the supplier in good time in writing whenever there is a change in the requirements specified for the items.

3.2 The supplier will inform the purchaser of any modification in the agreed system or in the quality assurance procedures, as well as any change in materials, production methods, subcontracted parts, datasheets and other documents, unless the supplier can positively preclude any detrimental influence of such a change on the properties and reliability of the items. This information will be provided sufficiently early and in such completeness as to permit the purchaser to review it and object to it before any modification is applied to the items. Silence on the part of the purchaser will not discharge the supplier from his sole responsibility as to the properties and reliabilty of the items.

3.3 If when testing items (2.1 and 2.2), the supplier finds an increase in detrimental deviations in the properties or reliability of the items with respect to the requirements specified in Annexe 1 of this agreement (impairment of quality), it will immediately inform the purchaser and take corrective actions such as improved production methods, materials, parts, test procedures, test facilities, etc. to ensure permanent remedies. Until these corrective actions take effect, the purchaser may request special measures (e.g. greater test frequency) for a reasonable period of time. Extra costs arising as a result of this will be borne by the supplier, unless the quality impairment can be shown to have been caused by the purchaser.

3.4 If the purchaser learns that any of the items do not meet the requirements specified in Annexe 1 of this agreement, he will inform the supplier without undue delay.

4 **Receiving inspection**

4.1 As section 2.2 above requires the necessary tests to be performed exclusively at the supplier's, the purchaser will only inspect the items on delivery for their class of goods and for any visible external damage to the packing which may have occured in transit.

4.2 In all other respects, the purchaser is released from the duty to examine for defects on receipt of goods.

5 **Validity**

This agreement will come into force on signature and may be terminated at the end of a calendar quarter on three months' notice. The notice of termination shall be given in writing. It will apply to all deliveries of items ordered after this agreement came into effect and for which the orders were acknowledged prior to the termination of this agreement.

6 **Arbitration**

All disputes arising shall be settled under the rules of conciliation and arbitration of the International Chamber of Commerce in Paris by three arbitrators appointed in accordance with the said rules. The place of arbitration shall be Geneva, Switzerland. The procedural law of this place shall apply where the rules are silent.

APPENDIX I
General quality specification

I.1 GENERAL

I.1.1 Scope

This quality specification establishes general quality requirements which are to be observed for the purchasing and delivery of components. It is binding for the customer and supplier when it forms a constituent part of purchase agreement or supply contract. It is valid in conjunction with datasheets, detail or family specifications which describe and define the function and the electrical and mechanical parameters.

It defines qualification requirements and serves as a guideline for approval procedures. Furthermore, it can serve as a foundation for a quality convention between the supplier and the customer.

I.1.2 Precedence

In the event of a conflict between requirements following order of precedence will apply:

1. purchase agreement or supply contract;
2. the detail or part specification of the customer or a manufacturer's datasheet which has been approved by the customer;
3. the customer's family specification;
4. this quality specification;
5. the latest edition of manufacturer's published datasheets; and
6. standards which are referred to in parts of this specification. The latest revisions at the time of placing the order are valid.

I.1.3 Vendor addendum

Any deviation from this quality specification must be mutually agreed by vendor and purchaser. These deviations must be stated in a vendor addendum.

I.1.4 Quality relationship

The general relationship between the customer and the supplying manufacturer is schematically shown in the flow chart of Fig. I.1.

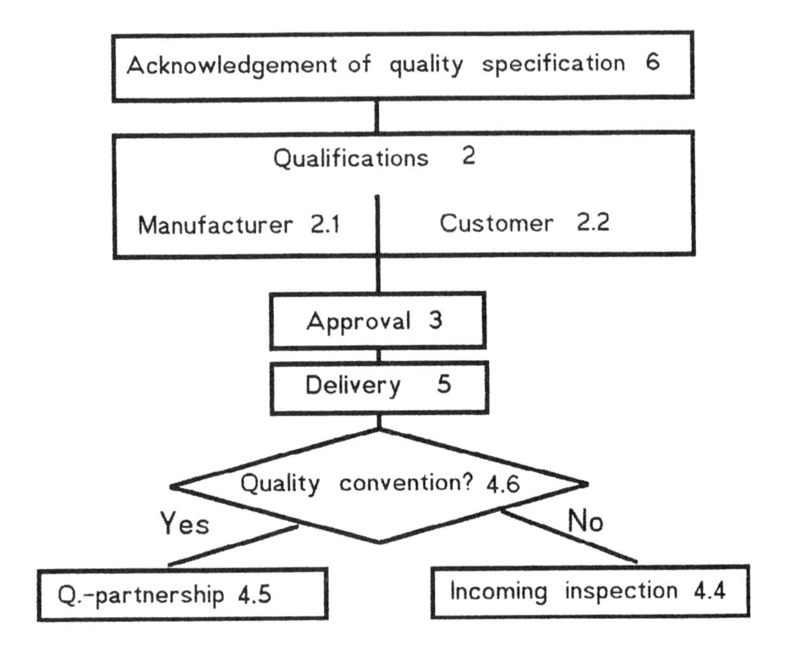

Figure I.1 Flowchart of quality relationship.

I.2 QUALIFICATION

I.2.1 Manufacturer's qualification system

It is necessary that the manufacturer maintains a qualification system capable of proving that the requirements of this specification are met. This is important for:

- functional and electrical characteristics;
- mechanical charateristics;
- reliability and environmental requirements; and
- manufacturing and processing feasability.

The reliabilty and environmental requirements can be met by qualifying representative types of a component family.

Manufacturer's requalification and monitoring

It is necessary that the manufacturer requalifies its products regularly at least once a year and that it maintains a quality monitoring programme for reliability and environmental tests to ensure and demonstrate that the delivered components meet all the agreed requirements.

Manufacturer's documentation and reporting

The supplier will submit to the customer the quality manual including the flowchart for quality assurance, applied sampling plans with accept/reject criteria, and other reports such as the status of qualification samples, qualification report, periodical requalification reports, once a year, and periodical monitoring reports, twice a year.

Periodical reports must include the results of failure rates (infant mortality), moisture endurance and operating life tests.

The manufacturer is obliged to indicate materials for which the regulation for dangerous goods and materials is applicable.

The documents must be retained for at least five years.

I.2.2 Customer's qualification

The customer reserves the right for itself, or a delegate, to perform tests according to the requirements of this specification. The qualification samples ordered by the customer must meet the following conditions.

1. They must originate from a production line of which the products have already been completely qualified by the manufacturer himself, including positive results of moisture endurance and life tests.
2. The samples must be recently manufactured and provided from a continuously running production line on which later deliveries of the relevant device will be manufactured.
3. They will not be specially prepared, tested and prescreened, unless such procedures are part of normal production flow.

4. A written statement will be given with the samples, confirming that the qualification samples are representative of future production for a period of at least 12 months.

I.2.3 Test procedures and requirements

Test procedures and requirements for reliability and environmental qualification are specified in Fig. I.2.

Test group	subgroup	Test procedure	MIL-STD 883	CECC 90000	Test conditions	acceptce. size	LTPD
0.1		Electrical Test			Applicable specification	100%	
0.2	B1	Physical dimens. coplanarity	2016	4.3	Applicable specification lead spacing	0/13	
1.1		External visual inspection	2009	4.2.2	MIL-STD-883 method 2009	0/22	10
1.2	B2	Resistance to solvents	2015	4.4			
1.3	B4	Internal visual inspection			Applicable specification	0/5	
1.4		X-ray analysis			Applicable specification	0/5	
2.1	B3	Solderability			Non SMD: Wetting: T=235° C, t=3s Dewetting: T=260° C, t=10s SMD: Wetting: T=215° C, t=3s Dewetting: T=260° C, t=30s		
		Visual examination			Reject if less than 95% covered surface		
3		Thermal sequence				0/32	7
3.1		solder preconditioning					
3.2		Thermal shock	1011	4.6.8	B: 15x,-55°C/125°C liquid		
3.3		Temparature cycles	1010	4.6.8	C: 100x, -65°C/150°C air		
3.4		Moisture endurance	1004	4.6.3	10x: 25°C...65 C/90% r.H.		
3.5		Visual examination	1010	4.2.2			
3.6		Electrical test			Applicable specification		
		Cracking sensitivity			Soak and soldering process simulation		
4	D4	Mechanical sequence			not applicable for plastic packages		
5	D2	Lead integrity sequence				0/22	10
5.1		Lead fatigue	2004	4.6.12	A and B2		
5.2		Visual examination	1010	4.2.2			
5.3		Adhesion of lead finish	2025		3 devices tested		10
6	D5	Salt atmosphere	1009	4.6.14	24h, 35°C	0/22	10
6.1		Visual examination	1009	4.2.2			
7		Moisture endurance solder preconditioning					
7.1		THB standard test		4.6.3	85°C, 85% r.H., t=2000h	0/45	5
		Electrical test			Applicable specification		
		instead of THB test: HAST test		4.6.3	130°C, 85% r.H., t=200h	0/45	5
		instead of THB test: Pressure cooker test			121°C, 100% r.H., 96h, without bias	0/32	7

Figure I.2 Test procedures and requirements.

Test group	subgroup	Test procedure	MIL-STD 883	CECC 90000	Test conditions	acceptce. size	LTPD
8		Life tests					
		Solder preconditioning					
8.1	C1	High temperature life test	1006	90100 4.2.2	Dynamic operation, Vcc=Vcc_max 125° C, 0....2000h or 150° C, 0...1000h	0/116	2
		Electrical test			Applicable specification		
8.2		Low temperature life test	1005	90100 4.2.2	Dynamic operation, Vcc=Vcc_max -20°C, 0....2000h the most sensitive pin combination must be included	0/116	
		Electrical test			Applicable specification, no param.degrad.		
9		ESD sensitivity					
9.1	B8	Human body model or	3015	90100 4.1.9	1kΩ, 100pF, 5 discharges, both polarities, V>= 1000V, the most sensitive pin combination must be included	0/11	
9.2		Charges device model			<1Ω, Device Cd, 5 discharges, both pol., V>= 500V		
		Electrical test			Applicable specification, no param.degrad.		
10		Temperature storage			Tj_max, 168h, for programmable devices	0/116	2
11		Photosensitivity test			Applicable specification	0/11	20
12		Latch-up (CMOS only)			EIA JEDEC Std. No. 17 $I_{trigger}$ ± 80 mA, V_{over} = 1.5*V_{max}	0/15	
13		Flammability Needle flame test or UL94,V0+ASTM D2863-77			IEC695 T.2.2, 20s heat, 15s extinguish	0/11	
14.1	B3	Bond strength test	2011		performed before moulding	4IC's	10
14.2		Die shear test	2019			0/3	

Figure I.2 Test procedures and requirements (cont'd).

I.3 REQUIREMENTS FOR APPROVAL

I.3.1 Product and process approval

The approval of the product and the process by the customer is a prerequisite for the ordering and delivery of components. To this end, the manufacturer provides the customer with all necessary information in the form of specifications and flowcharts containing production steps, in-line inspections, internal visual inspections, sampling plans, quality assurance and acceptance criteria, product screening and preconditioning procedures.

Approval by the customer is dependent on successful completion of the design, workmanship, electrical, environmental and reliability evaluations of products and processes submitted for qualification by:

- acknowledgement of the manufacturer's qualification; and
- successful termination of the customer's qualification.

I.3.2 Manufacturer's reporting duty

Product and process change notification

After qualification, the approved manufacturer may not carry out any significant alterations of the product or process affecting the shape, suitability, function processing or reliability without previously informing the customer.

The customer must be informed six months before the planned commencement of delivery, and must receive a description of the alteration. Such notification will include results of reliability and qualification tests with evidence showing that the performance, reliability and quality of the ICs continue to meet the specified requirements. The following changes require notification for ICs:

- mechanical dimensions
- marking
- package material, moulding compound
- packing
- lead material and coating
- die attachment method
- bonding material and method
- wafer size
- chip size or active chip area
- chip layout
- technology
- functional and electrical parameters
- subcontracting by manufacturer
- manufacturing location
- datasheet.

The latest layout photograph and identification number, as well as moulding compound, have to be submitted to the customer for qualification.

The customer will notify acceptance or rejection not later than six months after receiving the representative changed samples.

The manufacturer's reporting duty applies in any case where materials are introduced to which the regulation for dangerous goods and materials applies.

Problem reporting

After approval, the manufacturer is obliged to inform the customer without delay of problems which could influence the quality, the reliabilty, the processing or the usability of components already delivered.

Product discontinuation

If a manufacturer intends to discontinue the production and delivery of an approved product, the customer must be informed in time to make an adequate final order. The manufacturer must therefore deliver the following information at least 12 months before the end of production:

- date of final order
- date of final delivery
- suggestions for substitutes.

Withdrawal or change of safety approval or labelling

The customer must be informed immediately if an alteration of regulations or other reasons mean that the requirements for a safety approval or safety label are no longer met or are changed.

Mailing of new datasheets or data books

The publication of new datasheets or books does not fulfil the manufacturer's reporting duty according to the other stipulations given in this section.

I.3.3 Customer's privileges

Withdrawal of approval

The customer reserves the right to withdraw a given approval if violations of the specification are proven, which could influence the quality, reliability, processing or usability of the components.

The customer assumes that, in the case of a violation on the part of the manufacturer, the manufacturer will accept an order reduction or cancellation, or credit the returned goods.

A withdrawal of approval or prohibition can be pronounced in the following cases:

- gross defects in manufacture;
- unsatisfactory reliability;
- non-compliance, or alteration without prior agreement, of specified parameters;
- discovery of faults in a quality audit; and
- failing of a requalification.

Changes of specification

The customer reserves the right to change specification. Changes will be agreed with the manufacturer. The manufacturer will be granted an appropriate length of time for the introduction of possible changes.

Quality audit

The customer is granted the right to conduct an audit of the manufacturer's production and quality system, in accordance with international standards, during the course of a qualification as well as after an approval. For this reason, customer's representatives will be permitted to visit the manufacturer's plant supplying the components for qualification. The visits will be negotiated and scheduled in advance. The findings resulting from these visits will influence the approval or withdrawal decisions.

The customer will not make any demands which contradict the proprietary interests of the manufacturer.

I.4 QUALITY AND RELIABILITY

I.4.1 Manufacturer's internal visual inspection

This inspection has to be performed lot by lot prior to encapsulation or sealing, if applicable, in accordance with the valid regulations. For ICs, these are MIL-STD 883, method 2010B; AQL = 1 cumulative.

I.4.2 Failure rates

The manufacturer must prove failure rates for each component family based on operational life tests. The calculation relates to $T_{(ambient)}$ = 55°C, 60% UCL and an activation energy of 0.4 eV. The following values, which are valid for ICs, serve as a standard gauge while no other values are agreed upon.

Type	*Failure rate (fit)*
EPLD, FPGA	300
Asic (>10k gate functions)	200
Microprocessor	150
Microcontroller, EEPROM	125
Telecom circuits, EPROM, PAL, asic(<10k), VLSI	100
SRAM	75
DRAM	50
Interface, LSI	20
SSI, MSI (TTL, CMOS, linear)	5

If the infant mortality rate of a component exceeds the specified value the manufacturer must carry out an appropriate preconditioning (burn-in).

I.4.3 Incoming quality

AQL values

The manufacturer is obliged to ensure that the components delivered meet the agreeed specification. The following values serve as a gauge for the AQL of ICs:

Defectives		*AQL*
Visual and mechanical		0.1
Electrical	SSI, MSI	0.04
	LSI, VLSI, memories	0.065
	ULSI	0.1
	GLSI	0.15
Solderability		0.4

AOQ values

The manufacturer is obliged to prove that the components which are delivered meet the specified AOQ values. It is assumed that the values measured by the customer at the incoming inspection do not exceed the specified AOQ values.The following values can serve as a gauge for the AOQ of ICs:

Type	*AOQ (dpm)*
SSI, MSI, LSI	50
VLSI, ULSI	100
GSLI	150

Correlation

In order to comply with the requirements of this specification, the measurement values and methods which are used to prove the delivered quality and reliability must be correlated between the manufacturer and customer.

Retention of data

All documents concerning process and quality monitoring must be retained at least five years if not otherwise stated.

I.4.4 Incoming inspection

An incoming inspection may be performed by the customer as a sampling inspection or a 100% inspection. If the AQL value is exceeded in a sampling inspection then the customer is entitled to return the entire delivery lot.

If the manufacturer is not able to provide a replacement delivery for a rejected lot in good time, the customer can perform a 100% test on the rejected lot in agreement with the manufacturer. The costs of this 100% test will be borne by the manufacturer.

I.4.5 Ship-to-stock/ship-to-line

An incoming inspection by the customer may be omitted if the necessary conditions are met and a ship-to-stock or ship-to-line agreement has been reached between the manufacturer and customer. The following points must be defined in such an agreement:

- manufacturer's testing technology and quality assurance system
- manufacturing locations (wafer, assembly, final testing)
- AQL and AOQ values
- process control data
- minimum size of lots for delivery
- methods of dealing with problems
- all legal relationships including arbitration.

I.4.6 Quality convention

An intensive technical co-operation between the manufacturer and customer is necessary to achieve the aims of this specification. The foundations of such co-operation can be laid down in a quality convention. Satisfactory quality and reliability results in current deliveries are prerequisites for a quality convention.

Essential conditions for a quality convention are:

- a successsful zero-defect programme by the manufacturer
- outstanding dpm values
- outstanding fit values
- a successful ship-to-stock programme by the manufacturer
- punctual deliveries.

I.5 DELIVERY

I.5.1 Date of manufacture

The date code, i.e. the date of final testing, may not be older than one year. It is a prerequisite that the final testing takes place directly after manufacture. Components with a date code older than one year may only be delivered after a prior approval by the customer.

I.5.2 Marking

Components must bear the following marking, as far as size permits:

- type and/or ordering designation
- manufacturer's name or logo
- date of manufacture or date code
- place of manufacture
- pin number or polarity marking as specified
- special marking for selection or precondition if specified.

Marking must be clearly legible, indelible and resistant. If the size of the component is inadequate then the marking must appear on the packaging unit.

I.5.3 Packaging

The packaging materials must be free from emission and secretion, and must be suitable for use in ESD-protected areas, i.e. the surface resistivity must be less than $10^{10}\,\Omega$/square. In addition, for ESD-sensitive devices, the surface resistivity must be greater than $10^{5}\,\Omega$/square.

I.6 ACKNOWLEDGEMENT

This specification must be recognized as legally valid signed by the manufacturer's sales and quality departments.

Glossary

This glossary lists the definitions of terms used in the context of this book. A vocabulary can be found in ISO 8402.

Asic
: Application specific integrated circuit. The logic function of an asic is defined by the user to suit to his needs.

ATE
: Automated test equipment. Universally usable test equipment, software controlled by a ⇒ test program.

Audit
: Examination of a production facility to verify its compliance to a specified quality level.

Built-in self test (BIST)
: A ⇒ device or ⇒ component which tests itself with a self-generated test pattern and which gives the test result as a pass/fail condition.

Bus contention
: Fail functions of a bus line which is erroneously driven by several outputs at the same time.

Component
: In this context, a component designates the smallest structural member bought and/or used by the manufacturer of an electronic ⇒ device.

Coplanarity
: Worst case distance lead to pad of a ⇒ surface mounted package.

Customer
: Purchaser and ultimate user of a component.

Defect
: Imperfection of the physical structure of a ⇒ component, originated during manufacturing or later. A defect may become a ⇒ failure if the external behaviour of the component has deteriorated.

Device

In this context, an electronic device designates an electronic product manufactured by the purchaser of components. Elsewhere, device may be used as a synomym of component.

ECL

Emitter coupled logic, a technology used for high speed applications instead of conventional CMOS or TTL logic. CML, current mode logic, is a similar technology.

EOS

Electrical overstress, caused by inadmissible high voltage or current.

ESD

Electrostatic sensitive device or electrostatic discharge.

ESIS

Evaluation survey and information system; database which stores the evaluation status and results for all components.

Evaluation

Test activities to ascertain all essential properties of a new component.

Failure

Imperfection of the physical structure of a component which leads to an inability to perform its correct function. It may be caused by a ⇒defect or by mishandling.

Fault simulation

Simulation of the behaviour of a logic circuit in the presence of faults.

Functional test program generation

Method to generate ⇒test programs using a model based on the external functions of the component under test (black box model).

Ground(/earth) bounce

Noise appearing at the output of a circuit generated by other switching circuits sharing the ground line.

Gull wing lead shape
Leads of a surface mountable package which are bent outward. The contrary is a J lead package, where the leads are bent inward.

Infant mortality
Early failures at the beginning of the life time of a component. They can be screened out by preconditioning (like burn-in, thermal cycling or vibration).

Insertion mounted package
Package with pins, mounted on printed boards by through-holes e.g. DIP (dual in line package), SIP (single in line package).

Joint qualification
Procedure for sharing the qualification costs of a component between the vendor and customer.

Logic spike
Spurious signal at the output of a circuit caused by internal race conditions.

Manufacturability
Property of a component which permits its use in the customer's production line.

OEM
Original equipment manufacturer. Subcontractor from which complete subunits are bought either off-the-shelf or as an extended workbench.

PAL
Programmable array logic, belongs to the user programmable application specific circuits, together with GAL (generic array logic), PLA (programmable logic array), and EPROM (electrically programmable read only memory).

Program generator
Software which generates a test program for an ⇒ ATE system, taking into consideration all formal and electrical restrictions set up by the test system.

Qualification
All activities necessary to decide whether a component can be approved for usage.

Quality
Compliance of the behaviour, the failure rate and the lifetime of a product to the expectations of the customers.

Quality assurance
All planned and systematic actions in the whole production process necessary to provide adequate confidence that a product will satisfy given requirements.

Reliability
Probabilty that a component or device will perform its intended function for a given period of time under a given set of environmental conditions.

Requalification
Qualification repeated after a major change.

Seating plane
Plane of a printed board where the pads are located.

Structural test program generation
Method for generating ⇒ test programs using a model based on the internal structure of the component under test (gate model).

Surface mounted package
Leaded package which is mounted on printed boards by soldering on the surface, e.g. FP (flat pack), PLCC (plastic leaded chip carrier).

Test coverage
The percentage of all possible failures detected by testing a component. Usually only failures which are contained in a failure catalogue are considered.

Test program
Software to control an automatic component or board tester.

Threshold time

Time allowed for the input of a digital circuit to be in the active region of the transfer characteristics without causing malfunctions.

Vendor

In this context, the expression vendor means manufacturer of components.

Wear-out

Increasing failure rate at the end of the reliability life curve.

Further reading

Beside recent papers some historical papers are cited.

Accumolli, A. (1993) Analysis of run-in process for PC manufacturing. *Quality and Reliability Engineering International*, **9**, 407–10.

Barber, M.R. and Satre, W.I. (1987) Timing accuracy in modern ATE. *IEEE Design and Test*, April.

Bardell, P.H., McAnney, W.H. and Savir, J. (1987) *Built-in Test for VLSI*, John Wiley & Sons, Chichester.

Blanks, H.S. (1994) Quality and reliability research into the next century. *Quality and Reliability Engineering International*, **10**. March, 179–84.

Brombacher, A.C., Van Geest, E., Arenson, R., Van Steenwijk, A. and Herrmann, O. (1993) Simulation, a tool for designing-in reliability. *Quality and Reliabilty Engineering International*, **9**, 239–49.

Brombacher, A.C. (1992) *Reliability by Design*, John Wiley & Sons, Chichester.

Burggraf, P.S. (1984) The roles and use of failure analysis service. *Semiconductor International*, Sept., 43–9.

Bursky, D. (1993) Asic family crams up to 1.2M usable gates/chip. *Electronic Design,* June, 111–5.

Catt, I. (1966) Time loss through gating of asynchronous logic signal pulses. *IEEE Transactions Elect. Comp.*, Feb., 108–11.

Chaney, T. (1983) Measured flipflop responses to marginal triggering. *IEEE Transactions on Computers*, Dec.

Comerford, R. (1989) Comparing design verification systems. *Test & Measurement World*, Sept., 99–106.

Coppola, A. (1994) What's new in reliability engineering. *Quality and Reliability Engineering International*, **10**, March, 175–7.

Croes, R. and Hendrics, P. (1965) Standardizing latch-up immunity tests for HCMOS circuits. *Electronic Components and Applications*, **3**.

D'Heurle, F.M. Electromigration and failure in electronics: an introduction. *Proceedings of the IEEE*, **59**, Oct., 1409–18.

Dale, B.D. (1993) The key features of Japanese total quality control. *Quality and Reliability Engineering International*, **9**, 169–78.

DeFalco, J.A. (1970) Reflection and crosstalk in logic circuit interconnections. *IEEE Spectrum*, **9**, July, 44–50.

Dike, C. and Burton, T. (1989) Measuring metastability. *Signetics Application Note*, Nov.

Eichelberger, E.B. and Williams, T.W. (1973) A logic design structure for LSI testability. *Journal of Design Automation Fault-tolerant Computer*, **2**, 165–78.

Elsen, S.H. and Followell, R.F. (1993) The total cost of quality: what else should be contemplated. *Quality and Reliability Engineering International*, **9**, 203–208.

Flaherty, J.M. (1993a) New SMDs are leading us on. *Test & Measurement World*, Jan., 50–4.

Flaherty, J.M. (1993b) Choosing an environmental test-lab. *Test & Measurement World*, April, 39–40.

Flaherty, J.M. (1993c) X-rays stay on the leading edge. *Test & Measurement World*, April, 57–62.

Flaherty, J.M. (1993d) A Burnin's issue: IC complexity. *Test & Measurement World*, Oct., 61–4.

Fleischhammer, W. (1991) Good quality at low cost: trends in quality assurance. *Quality and Reliability Engineering International*, **6**, 485–8.

Fleischhammer, W. and Doertok, O. (1979) The anomalous behaviour of flip-flops in synchronizer circuits. *IEEE Transactions on Computers*, March, 273–6.

Frank, E.H. and Sproull, R.F. (1981) Testing and debugging custom integrated circuits. *Computing Survey*, Dec.

Gerling, W. (1990) Modern reliabilty assurance of integrated circuits, European Symposium on Reliability of Electronic Devices, Failure Physics and Analysis.

Goh, T.N. (1993) Taguchi methods: some technical, cultural and pedagogical perspectives. *Quality and Reliability International*, **9**, 185–202.

Gruetzner, M. and Starke, C. (1993) Experience with biased random pattern generation to meet the demand for a high quality BIST. *European Test Conference*, 408–17.

Hasseloff, E. (1993) Metastable response in digital circuits. Texas Instruments Application Note, EB204E.

Hayes, J.P. (1988) Detecting pattern sensitive faults in RAMs. *IEEE Transactions on Computers*, March.

Hnatek, E.R. (1975) User's tests, not data sheets, assure IC perform. *Electronics*, Nov., 108–13.

Hnatek, E.R. (1992) ISO 9000 for the test engineer. *Test & Measurement World*, Oct. Testing Supplement, 7–11.

Hnatek, E.R. (1993) The step toward ISO 9000 registration. *Test & Measurement World*, Feb., 89–92.

Ishikawa, K. (1985) *What is Total Quality Control*? Prentice-Hall Inc., Englewood Cliffs, NJ.

Johnson, B. (1993) Boundary scan eases test of new technologies. *Test & Measurement Europe*, Autumn.

Jones, K. (1993) Characterizing IC packages and interconnects. *Test & Measurement World*, June, 55–60.

Kacprzak, T. (1987) Analysis of metastable operation in SR CMOS flipflop. *IEEE Journal of Solid State Circuits*, Feb., 57.

Kirstein, H. (1987) How to start with a quality program. *EOQC Quality*, **2**.

Lapin, M. (1992) Testing in all three axes. *Test & Measurement World*, Sept. 87–93.

Lombardi, F. (1988) *Testing and Diagnosis of VLSI and ULSI*, Kluwer Academic Publishers.

Luecke, G. (1964) Noise margins in digital integrated circuits. *Proceedings of IEEE*, Dec., 1565–71.

Markowitz, M. (1992) Design for test without really trying. *EDN*, Feb., 113–22.

Marshall, M. (1993) How well does your supplier test asics? *Test & Measurment World*, Jan., 41–2.

Masteller, R.S. (1991) Design a digital synchronizer with a low metastable failure rate. *EDN*, April, 169–74.

Miles, T.E. (1972) Schottky TTL vs ECL for high speed logic. *Computer Design*, Oct., 79–85.

Mizko, A. (1986) *Digital Logic Testing and Simulation*, Harper and Row, New York.

Narud, J.A. and Meyer, C.S. (1964) Characterisation of integrated logic circuits. *Proceedings of IEEE*, Dec., 1551–64.

Nelson, N. *Accelerated Testing*, John Wiley & Sons, Chichester.

O'Connor, P.D. (1993) Quality and reliability: illusions and realities. *Quality and Reality Engineering International*, **9**, 163–8.

Pantic, D. (1986) Benefit of integrated circuit burn-in. *IEEE Transactions on Reliability*, April.

Parker, K.P. (1992) *The Boundary Scan Handbook*, Kluwer Academic Publisher.

Raheja, D.G. (1991) *Assurance Technologies, Principles and Practices*, McGraw-Hill.

Ratford, V. (1992) Boundary scan improves ATPG performance. *Test & Measurement World*, May.

Reghbati, H.K. (1991) *VLSI Testing and Validation Techniques*, IEEE Computer Society Press.

Richardson, A.M. and Dorey, A.P. (1992) Supply current monitoring in CMOS circuits for reliability prediction and test. *Quality and Reliability Engineering International*, **8**, 543–8.

Romanchik, D. (1992a) Burn-in: still a hot topic. *Test & Measurement World*, Jan., 51–4.

Romanchik, D. (1992b) Performing ESD audits. *Test & Measurement World*, June, 79–84.

Romanchik, D. (1992c) Overcoming the limitations of fault simulation. *Test & Measurement World*, Sept., 71–4.

Romanchik, D. (1992d) IDDQ testing makes a comeback. *Test & Measurement World*, Oct., 53–8.

Savage, R.M., Park, H.S. and Dr Fan, M.S. (1993) Automated inspection of solder joints for surface mount technology. *NASA Technical Memorandum 104580.*

Scheiber, S.F. (1969) The new approach to environmental test. *Test & Measurement World*, Sept./Oct., 71–84.

Scheiber, S.F. (1992) Evaluating test-strategy alternatives. *Test & Measurement World*, April, 57–60.

Schiessler, G., Spivak C. and Davidson, S. (1991) IDDQ test results on a digital CMOS IC. *Proceedings from Custom Integrated Circuit Conference*, IEEE.

Seichter, W. (1992) *ASET 600 Alpha Testsystem*, Siemens Nixdorf AG, Aügsbürg, Germany.

Shear, D. (1992) Exorcise metastability from your design. *EDN*, Dec., 58–64.

Singleton, R.S. (1968) The analysis of reflections in lossless transmission lines. *Electronics*, Oct., 37–9.

Smith, W.B. (1993) Total customer satisfaction as a business strategy. *Quality and Reliability Engineering International*, 49–53.

Soden, J.M., Charles, F.H., Ravi, K.G. and Weiwei, M. (1992) IDDQ testing: a review. *Journal of Electronic Testing*.

Steward, J.H. (1977) Future testing of large LSI circuit cards. *Proceedings of Semiconductor Test Symposium*, 6–17.

Stover, A.C. (1984) *ATE: Automated Test Equipment*. McGraw-Hill.

Tanaka, Y. (1992) *Perspective of Semiconductor Devices*, NEC Device Technology International, **25**.

Tannouri, F. and McCaffrey, J. (1993) Modified IC handlers reduce ESD failures. *Test & Measurement World*, Sept., 77–80

Totton, K.A.E. (1985) Review of built-in test methodologies for gate arrays. *IEEE Proceedings*, March/April, 121–9.

Wakerly, J. (1987) Designers guide to synchronizers and metastability. *Microprocessor Report*, Sept./Oct., 4.

Whittle, S., Smith, S., Tranfield, D. and Fester, M. (1992) Implementing total quality. *International Journal of Technology Management*, **7**.

Williams, T.W. and Parker, K.P. (1983) Design for testability, a survey. *Proceedings of IEEE*, Jan.

Yon-In, S.S. (1992) Maintain signal integrity at high digital speeds. *ED*, **77**, May, 77–92.

INDEX

Accelerated life test 9 123
Accuracy 173
Active loads 279
Algorithmic pattern generators 279
Analysis by analogy 290
ANSI/ASQC Q90 standard 250
AOQ values 310
Application failure 188
Application tests 285
Appraisal of failures 156
AQL values 309
Arrays
  cell 93
  customer-defined 93
  flex 93
  gate 93
  gate, second source for 108
Arrhenius equation 123, 136
Asics 35, 56, 63, 170, 313
  electrical evaluation 92
  full custom 92
  semicustom 93
  user programmable 111
Asparagus types 209
Asynchronous behaviour 78
  of flipflops 103
Audit 145, 313
Audit check list 146
Audit self-assessment 251
Auger electronic spectroscopy (AES) 213
Automated testing
  benefits of 62, 66, 229, 313
Automatic analysis of static characteristics 65
Automatic test equipment (ATE) 29, 63, 114, 168, 187, 279
  ATE data 27
  ATE system 172, 232, 265
  ATE test system 131, 149
Availability 13

Backdriving problem 182
Band gap 64
Bath tub curve 135
Benchtop testers 167, 171
Bergeron method 41
Bergeron method 254
Bit line failures 132
Board quality prediction 184
Board test 242
  management 179
  methods 181
  strategy 179
Board testers 181, 187
Board testing cost 185
Boundary scan 183, 192, 208, 279
Boundary scan technique 103
Brittle fracture 213
BS 5750 specification 250
Built-in self test (BIST) 103, 107, 184, 192, 208, 285, 313
Burn-in 135
Burn-in procedure 122
Bus contention 73
Butterfly pattern 178

Calculation of crosstalk 257
Calibration of test equipment 229
Capacitance of inputs and outputs 75
Capacitor discharge method 131

Cell arrays 93
Cell failures 132
Cell libraries 109
Chain noise immunity 45
Characteristics
 static 15
Charged device model (CDM) 125
Computer aided database 155, 234
Constant current method 131
Coplanarity 143, 222, 313
Corrective actions 156, 157
Corrective loop 162, 163, 189, 203
Correlation with the vendor 243
Cost comparison 66
Cost
 of board testing 185
 of failure detection 17
 of final testing 199, 237
 of maintenance 205
 of quality 8, 238
 of test 217
Coupling factors 260
Coupling length 70
Coupling resistance 293
Cracking phenomenon 144
Crosstalk 41
 calculation of 257
Crosstalk on interconnection lines 253
Current sense method 77
Customer audits 145
Customer's privileges 308
Customer-defined arrays 93

Data package 26
Datasheet values 21
Datasheets, vendor's 12
Defect rate 216
Delay 60
 parameters 53
 time 53
 pattern sensitivity 56
Delivery 312
Design
 for manufacturability (DFM) 241
 for quality (DFQ) 241
 for quality 241
 for reliability (DFR) 241
 for test 193
 for testability (DFT) 162 191
 for testability (DFT) 241
Design information, transferring options 111
Design tools 14
Design verification system (DVS) 168
Design, vendor optimized 109
Digital function test 182
Distortion, pulse width 56
Dynamic
 characteristics 15, 52
 of input/output cells 104
 evaluation test 152
 loading 75
 noise immunity 68
 parameters 114
 performance 101
 power supply current 72
 testing 178, 182

ECL circuits
 evaluation 115
 evaluation tests 115
 technology 45, 115
Electrical evaluation 35
 of asics 92
 of memories 114
Electrical overstress (EOS) 124

Electrical parameters 68
Electromagnetic coupling 198
Electromigration 75, 140
EMI 293
  failures 197
Emulation test 182
End test 194
Environmental data 27
Environmental tests 9, 108
EN 29000 specification 250
EOS 314
ESD 314
  audits 227
  handling 213
  immunity 15
  potential 129
  prevention 122, 224
  sensitivity 124, 188
Evaluation 39
  by automatic test systems 62
  costs and benefits 145
  data 21, 33
  data packages 22
  functional and electrical 35
  measurements 22
  of ECL circuits 115
  of passive components 116
  of standard LSI and VLSI circuits 86
  samples 30
  survey 27
  test program 265
  tests on ECL circuits 115
Evaluation survey and information system (ESIS) 29
Expert system 203, 204, 208

Failure
  analysis 209, 243
  clusters 163
  coverage 287
  detection 21, 155, 178, 181
  diagnosis 188, 190, 196
  prevention 21, 219
  rate 121, 180, 218, 267, 309
Failure detection 147
  costs of 17
Failure management system 154
Failure mode and effect analysis (FMEA) 233, 234
Family specification 248
Fault simulation 176, 314
Fault tolerance 39
Field programmable asics 111
Field-induced model 125
Final test 194, 243
  cost of 199, 237
Flex-arrays 93
Flipflops 103
Full custom asics 92
Functional evaluation 35
  method 288
  simulation 95
  tests 15, 266

Gate
  arrays 93
  count 1, 37
  density 1
  transfer 106
Golden device 285
Go–no go tests 167, 171
Ground bounce 73, 314
Ground noise 104, 112, 197

Hardware testing 166
Heisenberg uncertainty principle 79
High-temperature life test 123
Highly accelerated stress test (HAST) 123
Hot electron effect 140

Human body model (HBM) 124
Humidity test 123

ICT 192
In-circuit tester 182
Incoming inspection 30, 58, 165, 219, 238, 243, 365, 311
Incoming quality 309
Infant mortality 188, 315
Inflammability 16
Information System 27
Input/output (I/O) characteristics 40, 50
Input/output cells, characteristics of 104
Integrated quality assurance 9
Internal clock skew 104
ISO 9000 230
  registration 145
  certification 249

Joint evaluation 21
Joint qualification 6, 21, 97, 114, 115, 118, 121, 151
Just-in-time (JIT) delivery 244

Knowledge base 204

Lab data pack 74
Laboratory data 27
Latch-up effect 130
Latch-up sensitivity 131
Lead coverage 143
Lead surface material 143
Lead time 149
Leakage current 77
Level scan technology 203
Level-sensitive scan design (LSSD) 182, 191, 279
Life tests 16, 108
  accelerated 9
Loadboard 16, 30, 174
  performance 107
Localisation of failures 211
Logic
  design 241
  function 38, 101
  simulation 89
  simulation patterns 287
  spikes 60
LSI and VLSI circuits 86

Maintenance cost 205
Manufacturability 14
Manufacturing process control 157
Marching pattern 178
Marginal
  input condition 79, 81
  pulse width 81
  set-up time 81
Marking 312
Mean time between failure (MTBF) 12, 21, 5, 79, 83, 205, 232, 236
Mean time between service calls (MTBSC) 19
Mean time to repair (MTTR) 208
Memories, electrical evaluation of 114
Metastability 70, 79, 83, 91, 112, 116, 242
Metastable behaviour 151
Metastable states 79
Module test 193
Moore's law 1

Needle adaptor 182
Noise 60, 72
  immunity 45, 68
  ground 104, 112

Optical inspection 211
Optical test 181
Optimal scheduling of tests 201

Optimal test strategy 179

Package opening 211
Packages
  hermetically sealed 141
  plastic encapsulated 141
Packaging 312
Parameters, static 88
Part specification 248
Passive components, evaluation of 116
Pin mapping 279
Power supply current, dynamic 72
Power supply, OEM 118
Preconditioning 135, 224
Prescreening 181
Pressure cooker test 124
Process
  changes 232
  control 158
  qualification 233
Procuring test software 175
Product
  approval 305
  control 158
Program generator 175, 315
Propagation of signals 253
Pseudo-sporadic failures 197
Pulse width 81
  distortion 56
Purchase contract 249

Qualification 67, 242, 302, 316
  procedure 24
  reports 31
  tests 5, 230, 242
Quality actions 157, 159
Quality assurance agreement 297
  design phase 15
  production phase 17
  prototype and early production phase 16
  integrated 9
Quality
  audit 308
  convention 311
  costs 238
  culture 9
  data 155
  group 17
  management 153, 154
  monitoring 19, 245
  performance 155
  relationship 302
  specification 122, 247
  specification 24, 247, 301

Recovery voltage 44
Reflection method 77
Reliability 14
  data 27
  evaluation 121
  questionnaire 269
  tests 118, 123
Reporting duty 306
Requalification 26, 30, 63, 66, 316
Requirements for approval 305
Resolution 173
Resource per pin testers 168
Reverse engineering 212
Ring oscillators 101
Ringing 41, 43
Rise and fall time 58
Run-in 224

Scanning electron microscope (SEM) analysis 211
Schottky 41, 52, 70, 83
Second source for gate arrays 108
Selecting the right vendor 171

Self test 182
Self-assessment audit 251
Series gating 115
Service
  contracts 205
  processor 203
  requirements 207
Set-up time 81
Shared resource testers 168
Shielding 293
Ship-to-line 244, 311
  procedure 166
Ship-to-stock 177, 244, 311
  agreement 248
  contract 24, 178, 249
  delivery 218
  procedure 57, 89, 165, 180, 195
Simulated failures 287
Simulation
  data 27
  model 89
  software 14
  functional 95
  logic 89
Soft errors 131
Software
  database 94
  tools 106
Solderability 16
Soldering methods 118
Specifications 247, 248
Spice program 22, 254
Spikes
  logic 60
  supply 72
Sporadic failure 188
Standard cells 103
Standards 247
  BS 5750 250
  EN 29000 250
  ISO 9000 247
Static
  characterization 64
  leakage current 77
  parameters 114
  testing 182
Static characteristics 15, 39, 49
  automatic analysis of 65
  of input/output cells 104
Statistical process control (SPC) 220, 222
Structural method 286
Subcontractors (OEM) 118
Substitution test 182
Supply current characteristics 48
Supply spikes 72
Symbolic layout 95
Synchronizing failures 197
System test 184

Teamwork procedure 228
Temperature–humidity bias (THB) 123
Test
  by DFI 213
  capacity 170
  chip 97, 98, 116
  circuits 112
  cost per failure 217
  economy 162
  facilities 16
  flow 196
  handler 202
  methods 285
  of standard cells 103
  pattern 114, 287
  plan 161
  procedures 304
  program 16, 30, 175, 182, 265, 282, 314, 316
  requirements 304
  set-up 53

software 63, 285
Test equipment evaluation 172
Test labs 24
Test strategy 160
Testability 13, 107
Tester accuracies 174
Testing
dynamic parameters of asics 178
handling 16
hardware 166
Tests
functional 15, 266
under environmental stress 9
Thermal cycling 140, 224
Threshold time 70
Time to market 241
Timing analysis, static 95
Timing generators 282
Timing simulation 95
Total quality 17
Total quality control (TQM) 228
Total quality culture 11
Traceability 162, 166
Tracking system 163, 166
database 209
Transfer characteristics 45
Transfer gates 106
Transition time (rise and fall time) 58
Trend analysis 290

Valuation of detected failures 156
Verification of failures 211
Vibration screening 140, 224
Visual test 181

Waiting times 85
Wear-out 140, 317

X-ray analysis 211
X-ray transmission radiography 220